This book is dedicated to all astronauts

THE GENSTAR SAGA

Rob Alexander

JumpFish Publishing
Sydney

First published September 2025 by JumpFish Publishing.

This edition, September, 2025.

ISBN-13: 978-0-9945792-5-6

Copyright © 2025 JumpFish Publishing

The right of Robert Alexander to be identified as the author of this work has been asserted by him in accordance with the Copyright Design and Patents Act 1988.

9 8 7 6 5 4 3

A CIP catalogue record for this book is available from the Australian Library.

Printed and bound by Ingram Spark.

Acknowledgements

I extend my gratitude to Roland Gesthuizen for his scatter-brained ideas that inspired this story.

"Hence it may easily come about that an especially strong, and therefore particularly shut-off and uninfluenced complex, becomes an "excessively valued idea," i.e., it becomes a dominant, defying every criticism and enjoying complete autonomy, until finally it comes to be an uncontrollable factor ..."

– Carl Jung, *Psychological Types* (1921) –

– 1 –

ADRIFT

Mia Hawthorn pressed her palm against the viewport, the thick glass leaching the warmth from her skin. It was the first time in her life she had been this close to the vacuum of space. Outside, Eridani-B hung suspended in the void, a blue-green marble swelling imperceptibly with each breath. Sunlight—real sunlight from a star not yet their own—cast long shadows across the Landing Craft's cramped cabin, highlighting the copper tint in her hair and the fine stress lines at the corners of her eyes.

Sixteen thousand years of human history had led to this moment. Generations had lived and died aboard the GenStar, their lives confined to metal and polygrout-rendered walls, never knowing the caress of mountain wind or the splash of waves on naked skin—all so nine people could drift toward an alien world that might one day become humanity's second home.

Nine people out of two thousand five hundred.

Behind her, the soft whir of Charlie's servos broke the silence as the simulant moved through the cabin, checking life support readings with mechanical precision. His generations-old body, a patchwork of battered titanium panels dulled and scratched from centuries of use, clicked faintly at the joints. The worn silicone face—still pliable enough to crease into a faint, reassuring smile—sat above a Thunderbird-like mouth, its lower lip shifting as he spoke.

"Atmospheric entry in approximately nineteen hours," Charlie announced in that carefully calibrated tone designed to soothe human nerves. "All systems nominal."

Mia kept her eyes fixed on the starfield. "And our passengers?"

A calculated pause. "Navigator Vandergamma is in the pilot's seat. Anna Teoh is continuing her ... calculations. Dr. Chen is monitoring Callum after sedating him." The simulant's voice, usually neutral, carried a faint edge of judgment as his lower lip twitched slightly. Mia looked at him, considered arguing, but stopped herself. It was true, she had lied to Callum. Told him Rivel was at the Landing Craft, waiting, when she was already dead. It was a horrible thing to do, but it had saved his life. And right now, that mattered more than being a good person. She could live with that. What they could not live with was his outburst, the smashing of electrical panels, the animal rage. She'd had no choice but to order Chen to tranquilize him. The decision had settled the crew, finally freeing them from the chaos that had nearly swallowed their last chance at survival.

Mia turned slowly, her gaze settling on Anna. She was hunched over her tablet, tracing equations with her fingers, murmuring to herself or perhaps to unseen voices.

"And Devory?" Mia asked, not hiding the edge of concern in her voice.

"He's gone down to the cargo hold."

"Why?"

"He said it needed rearranging."

Mia shook her head, pushing away from the viewport. Her muscles ached, a bone-deep weariness that sleep could not cure. The emergency lighting bathed the cabin in a dim red haze, an unspoken warning that safety was not assured.

"And Phil and Park?" she asked after a moment.

"They're in the pilot's quarters. Dr. Chen informed me that Seong-Min Park is pregnant with Phil's child. Phil is providing comfort."

Mia exhaled slowly. So much for unsanctioned relationships. She wondered how many other rules they would break in the coming weeks, how many ancestral laws would unravel in the face of necessity. Too many.

She made her way to the small lab cubicle she had improvised from emergency medical supplies and jacked her data stick into a neural analyzer, one of the few pieces of equipment she had ordered during the last round of requisitions.

"Display results for Chen, Murchison, Vandergamma, Teoh, Broadbent, Park, Polsen, and Hawthorn," she commanded.

The screen flickered to life, showing nine neural patterns side by side. Eight human, one artificial; eight altered, one unchanged. This was the new baseline. Over the next few days, she would rescan everyone. She knew the patterns might transform, mutate. Callum in particular worried her, his grief already amplified by the Shard's distortions. His emotional state would almost certainly shift the outcome.

"You should rest," Charlie said, his silicone face creasing faintly into that familiar look of concern, lower lip dipping. "You have been awake for thirty-seven hours."

"I'll rest when we land," Mia replied, though they both knew it was a lie.

Charlie's camera eyes, glowing a warm amber as they tracked infrared and ultraviolet wavelengths, focused on her face, scanning, analyzing. "Your facial muscle tone suggests elevated cortisol levels. Your cognitive function is impaired by approximately twenty-three percent."

"Only twenty-three?" Mia said, attempting a smile that did not reach her eyes. "I am doing better than I thought."

The simulant did not respond to her gallows humor. Instead, Charlie extended a hand, palm up. In the center rested a small glass container, sealed with a simple plastic cap that seemed absurdly inadequate for what it held: a microscopic piece of the

Shard, barely visible to the naked eye.

"You requested hourly confirmation," Charlie said, his lower lip shifting with each syllable.

Mia looked at the container as Charlie held it up to the emergency lighting. The specimen caught the red glow, seeming to absorb and transform it into something darker, something that hurt her eyes to look at directly.

"Still there," she murmured. "Still real."

"The size is small enough to be almost innocuous," Charlie assured her, his silicone face steady. "The specimen remains stable."

But for how long? The question hung unspoken between them. How long before this tiny piece of something they could not understand changed them further? How long before one of them followed the same path as those they had left behind?

A soft thud from the cargo hold interrupted her thoughts. Devory. Probably rearranging the equipment again. She suspected it was an excuse to search for more pieces of the Shard, his obsession clearly returning.

"I should check on him," Mia said, turning toward the sound.

"Dr. Hawthorn." Charlie's voice stopped her. "May I ask a question?"

Mia paused, surprised. Charlie rarely initiated conversations. "Of course."

"Your foresight in bringing the neural data is going to prove invaluable over the coming days, but ... are we still ourselves?"

In any other circumstance, the question would have sounded out of place, even bizarre, easily dismissed with a wave of the hand. But this was no ordinary situation. And she was painfully aware that he had just torn away the fragile bandage she had pasted over her sense of self, leaving her exposed to all her fears. What could she say? That everything was ok? That the human mind would prevail no matter what was thrown at it? No. She had to be real. Gather herself, present at least the surface of sanity.

"I don't know," she admitted, telling the truth. "Our default mode networks are scrambled, but our core identities—fragile as they might be—seem to remain functional. At least for now."

"For now," Charlie repeated, his lower lip dipping slightly, the words carrying the weight of their uncertainty.

From the corner, Anna's voice rose suddenly, breaking the tension. "The probability curves are wrong," she announced to no one in particular, her eyes never leaving her tablet. "Reality should not bend that way."

Aldo's voice drifted back from the pilot's seat. "Anna, focus. We need those landing calculations."

"I am focused," Anna insisted, her voice taking on the dreamy quality that had grown more pronounced since their escape. "I am seeing all possible landings simultaneously. Quantum super-position of outcomes."

Mia and Charlie exchanged glances, his camera eyes flickering briefly to blue. Before the Shard, Anna had been the GenStar's most brilliant mathematician, precise and methodical. Now her genius had shifted, less tethered to conventional reality, seeing patterns and connections that should not exist.

"I'll help her," Mia said quietly to Charlie. "Keep monitoring the specimen. Any change, any change at all—"

"You will be notified immediately," Charlie confirmed, his silicone face creasing faintly with resolve.

As Mia crossed to Anna's corner, she felt the enormity of the situation press down on her, as if gravity had multiplied. They had survived when thousands had not, escaped the GenStar in a blur of desperation. If only Rivel had made it. Poor Rivel. She really did go down the rabbit hole. Still, Mia could not shake Rivel's last words: "Why do we feel everything is connected?" What had Rivel seen in those final moments? What had set all this in motion, turning their home into a graveyard, leaving only nine of them to drift towards an uncertain future?

– 2 –

10-MT DECELERATION BURN

The klaxon sounded three times—long, short, long—a signal etched into the Arkonauts' collective minds like a call to prayer. And like a call to prayer, the crew didn't rush or panic; they walked calmly—some briskly—to secure whatever loose items they could, tying them down, locking them away, or shelving them to ensure that nothing would become a projectile during the upcoming maneuver. This same ritual unfolded on every level of the onion-layered decks of the GenStar, a massive asteroid converted into a generation starship, turning it into a hive of activity. Children gripped their harnesses with a mixture of fear and excitement, while the veterans moved with steadfast efficiency, securing their own restraints only after checking the handiwork of others, some of whom nattered nervously about the usual forgetfulness that led to a smashed plate, or pot plant, and the inevitable cleanup that followed after the pulse had subsided.

From bow to stern, radiation shields slid into place with a soft hum, sealing the ship in a protective cocoon. A single chime over the PA confirmed full deployment, its soft tone momentarily interrupting conversations, reminding everyone that this was getting real.

On the Main Deck, Captain Frederick Carson stood rigid, hands clasped behind his back, eyes locked on a large monitor scrolling real-time data. His pulse remained steady, conditioned by decades of command, though a familiar tightness had built up between his shoulders and neck. In the upper right of the screen, 82 Eridani appeared as a well-defined disc, shimmering like a mirage. But Carson wasn't watching the star. His focus rested on the rotation value: 1.091 rpm. He knew it would wobble slightly during the shot, a subtle confirmation of its effectiveness. Up here, in the outer ring, the Arkonauts would experience it as a sudden compression of 8gs lasting a full second, their backs facing the direction of travel, since the ship had flipped around for its deceleration burn.

"Orbital insertion in 92 days, 14 hours, 37 minutes, 22 seconds," Aldo Vandergamma announced from his navigation console. He adjusted the holographic trajectory, aligning its parameters to Carson's view. His voice carried a faint tremor, an echo of the Pneumatic Blowout of 3991 that still haunted him and other survivors. "Today's shot will reduce velocity by 75.290 meters per second, maintaining approach vector."

Beside him sat his mathematical assistant, Anna Teoh, verifying the calculations with logical precision. She wore the guild pin on her collar not out of pride, but out of duty. Its fractal design reflected the essential nature of her role: that she was merely a self-repeating part of the whole. This was precisely why Aldo had selected her from among the candidates that had competed for her position. He recognized a mind that would prioritize duty over personal interests, a trait that was integral to the corps' ethos, which held that the collective journey of humanity required personnel who functioned as reliable components rather than unpredictable variables.

"Current velocity?" Carson asked, eyes still fixed on the rotation value.

"17.802 kilometers per second relative to 82 Eridani," Anna replied, her tone as measured as her calculations.

Carson nodded. "Proceed with the shot."

He placed his palm on the authorization panel, its surface worn from thousands of activations. The scanner verified his identity, linking him to the First Commanders who had launched the Gen-Star from the asteroid belt sixteen millennia ago. Originally of Martian descent, Carson's ancestors had weaker bones. However, after millennia of 1g artificial gravity his forebears had regained the robust skeleton of Earthlings, passing that trait down to him. This legacy not only granted him psychological authority but also the physical strength to command the ship.

In the Propulsion Control Sector, deep within the stern of the asteroid, Lionel Timmerman stood before a massive vault door. Hazard symbols in three languages—Ancient English, Middle Transit, and Modern Standard—marked its surface, each representing an era in the mission's long history. Heavy radiation shielding surrounded him, a necessary precaution for what lay beyond. The minimal gravity here would have left him floating if not for his magnetic boots, which anchored him with gentle clicks against the grated metal flooring. The air hung thick with the scent of ozone and heated electronics, dulled by the faint, stale odor of recycled atmosphere that had cycled through the filters too many times.

"Copy that. Awaiting secondary authorization," Lionel reported in his comms unit.

"Prometheus-Seven-Delta-Sundown," Carson responded.

Lionel entered the code into the keypad, its edges also polished by years of use. The vault door slid open with a low groan, revealing the Bomb Storeroom, a temperature-controlled chamber housing the GenStar's remaining thermonuclear devices. Cylindrical containers stood in precise formation, arranged according to a strict maintenance schedule run under his command. The room, once filled to capacity, now stood nearly two-thirds empty—a silent reminder of the journey's progress and the razor-thin margin for error in their remaining trajectory.

"Striker" Vasily Petrov entered the loop, the man officially responsible for the shot. It would ultimately be his hand that would trigger the blast. "Proceeding with shot 4,372. All systems stand by." He tapped his tablet in a nervous tattoo, a habit others found irritating but grudgingly acknowledged was unavoidable in these tense circumstances. Consulting his figures, he announced: "Ten-megaton package for 75.290 meters per second delta-v. Cradle Three prepared."

Lionel responded: "Roger, 10-MT," and approached a large container marked 10-MT: 11A-349. He activated the biometric lock and entered the access code, his movements deliberate despite having performed this procedure dozens of times. The container opened to reveal an unadorned cylindrical device, its simplicity belying its monstrous energy density. The diagnostic panel showed all parameters within acceptable ranges.

"Device nominal," he confirmed, checking the seal from the previous maintenance cycle. "Fusion package stable, explosive lenses and fission triggers within registered tolerance."

Vasily nodded. "Guess we'll find out when we blow it."

Lionel updated the status log. "Confirming device 11A-349." Verbalizing the serial number was a required formality, executed for the same reason that a doctor had to confirm a patient's name before amputating a leg. As he did so, he began operating the overhead block and tackle system, its chain gears clanking noisily as it descended. Countless refurbishments hadn't changed the basic mechanism. Some technologies, it seemed, were irreplaceable. Lionel secured the harness around the device, carefully positioning them on balance points marked on the casing. The nuclear package rose with glacial precision, whereupon he moved it along its track to Cradle Three.

After transferring it, the cradle's latches snugged with a solid click, instantly sending reference data to Propulsion Control.

Vasily verified the connections over his comms. "Cradle integrity confirmed."

"Initiating yield verification," Lionel announced, as he activated the cradle's integrated diagnostics panel. The system ran a final check of the bomb's fuel status and delivered the value: "Yield verified at 10.02 megatons. Nominal for navigation, confirm."

"Confirmed," Vasily replied.

This was his signal to haul it out of the storeroom. The rubber wheels of the trolley carrying the cradle rolled silently over the metal floor as his magnetic boots clickety-clacked in the mostly empty chamber. It was lonely in here. There was something about a blast that warmed things up and he looked forward to it.

Outside, the Firing Team was waiting for him. The "Stoker" took the trolley off him and approached the Bomb Tube, a conduit running through the asteroid's core. His lead apron shielded his body from the harsh radiation that emitted from the tube's interior. The dense material made his movements sluggish. Yet no one held this against him. The apron's surface was graffitied with the names of past Stokers, an informal memorial to those whose shortened lifespans were the cost of their critical role. This was compensated for by receiving the best of everything—food, water, living quarters—and, above all, respect.

"Cradle Three, ten-megaton package, confirmed," the Stoker said after checking his wrist display. "Tube pressure equalized. Path is clear."

"Proceed with insertion," Vasily directed.

The Stoker aligned the cradle with precision, his hands steady despite his weighty garment. The others stood back at a safe distance, watching and waiting while the tube's vacuum was replaced by ambient pressure.

On equalization, the Stoker opened the heavy reinforced tube hatch, pushed the cradle inside, so that it rolled off the trolley onto the tube's rail system, then closed the hatch again. After checking that the seal was tight, he tapped his wrist control, evacuating the tube so that it returned to vacuum. "Tube ready ... plunger in three, two, one ..." A humming began emanating from the tube. "Package in transit. Arrival at Blast Bell in 47 seconds."

Inside the Pneumatics Control Chamber, the Pneumatics Team Leader, Seong-Min Park, monitored the six piston shock absorbers attached to the Blast Bell's base. Her visor processed real-time data, giving her access to subtle pressure fluctuations that could spell the difference between success and failure. The pistons, pressurized to absorb the detonation force, could be lethal if containment failed, or if one of them was unbalanced. Above her station hung a memorial plaque for the Seventeen—those who had perished in the Blowout.

"Pneumatic Team, status report please."

Each specialist responded in sequence, as protocol demanded:

"Chambers one and two, optimal pressure and holding."

"Chambers three and four, optimal pressure and holding."

"Chambers five and six, optimal pressure and holding."

"Anchor points secure, strain gauges nominal."

"Latch mechanisms primed, override circuits functional."

Park delivered her final report to Vasily: "Air reserves for fifteen cycles captured. Filters at 97% efficiency. Containment tanks at near vacuum. Pistons set for 75.290 meters per second delta-v. We are go."

"Acknowledged," Vasily confirmed. "Package in transit. Prepare for compression."

"Copy, over."

"Thermals?" Vasily asked.

One of the newest members on the team, the officer in charge of the Thermal Management System, spoke up and said, "Heat sinks primed for absorption and dissipation. Coolant circulation at optimal flow."

"Excellent. Over to Arm."

The Robotic Arm Technician monitored the package from a wide-screen display. It gave him a clear view of the Blast Hatch at the base of the Blast Bell. The two-hundred-and-twenty-meter bell-shaped structure would direct the blast outward with the efficiency of a rocket nozzle. Below, a massive one-axis gimbal held the Bell steady, awaiting final vectoring.

"Package arriving ..."

The hatch opened. The cradle emerged. The Arm Technician guided it into place, securing it against the Bell's surface. "Cradle secured. Replacing hatch." Half a minute passed as the Bomb Tube was repressurized. "Arm Control evacuated."

Next, the Gimbal Technician read off the Bell's vectors according to the Chief Navigator's parameters. He monitored the hydraulics from a heads-up display and proceeded with the adjustment.

"Stand by for vectoring," he reported. The Bell moved slowly, driven by electric pumps that powered the hydraulic fluid. A steady thump-thump-thump emanated through the floor as the hydraulic pumps chugged. "Gimbal alignment complete and locked. Pneumatic integration secure."

Vasily tracked the process from the Propulsion Control Center, confirming: "All systems nominal. Detonation sequence ready." His display processed thousands of data points, flagging three minor anomalies—patterns he'd seen in previous burns. One concerned a water pressure valve in Cooling Loop 7, not mission critical, but Vasily knew better from the Blowout: small anomalies demanded attention. He issued an order to temporarily shut it down.

Lionel Timmerman had joined him by this stage. They stood together shoulder-to-shoulder in preparation for the final phase. Over the comms, Vasily said, "Proceeding with the countdown. All sections secure."

From the Main Deck, Carson's voice was crisp. "Copy."

Lionel took the arming key off the lanyard around his neck and held it up for Vasily to see.

"Initiating arming sequence," Vasily announced. He took the arming key off his own lanyard and inserted into the console and turned it 90 degrees. A panel slid open, revealing a second key slot. "Second key authorization ready."

Lionel Timmerman stepped forward and inserted his key into the second slot. "Second key confirmed."

"Arming sequence initiated," Vasily continued, entering a twelve-digit code. "Fail-safes disengaged. Detonation in T-minus 60 seconds."

The final klaxon blared. Arkonauts strapped into acceleration couches, some muttering secular prayers while others remained silent, observing the quiet of the Continuance.

Carson and his officers secured themselves, while the remainder of the Propulsion Team retreated to the "Bunker"—a reinforced space behind Propulsion Control. The air was heavy with the exhalations of nervous crew, recyclers humming faintly to keep pace.

"30 seconds," Vasily called from the Bunker. "10... 5... 4... 3... 2... 1... Detonation."

In an instant, an artificial sun was born. Its blinding flash penetrated rock and steel alike, imprinting X-ray visions of skeletons onto the back of eyeballs. Then came the WHUMP, a bone-liquefying punch that jarred teeth and scrambled brains. The colossal pneumatic pistons rammed inward, compressing two hundred meters of atmosphere into a supercritical fluid in under a second. The energy transferred directly into the GenStar's superstructure with a violent and unyielding shove. God's fist, some called it. It slammed the Arkonauts into their deceleration couches like pancakes, crushing the air from their lungs, blacking out the weak, cracking the bones of the frail.

Outside, if someone had watched from a separate spacecraft, they'd have seen dust shake off the asteroid's surface in a vast cloud. They'd have seen the behemoth emerge from its shroud, radio and telescope arrays vibrating like a struck gong. And lastly, they'd have seen the burn's fierce glow fade, leaving the GenStar's trajectory shifted back a notch, another small step on its descent into Eridani's gravity well.

And then it passed.

"Propulsion complete," Vasily responded, regaining his breath. "Systems nominal. Noting anomalies in Cooling Loop 5 and Piston Four for priority maintenance."

"Initiating post-detonation telemetry," Lionel added, activating specialized systems to analyze radiation signatures and detonation efficiency. "Neutron flux readings within expected parameters. Radiation decay pattern nominal. Ablation nominal. Initiating comprehensive detonation analysis."

"Compression nominal ..." The characteristic hiss of the compressed air releasing into the storage tanks reverberated through the superstructure. "All pistons recycled. Marking Piston Four for priority inspection," Park reported.

"Bell extension complete," stated the Gimbal Technician. The gimbal locked into its neutral position. Bearing pressures were noted, as well as heat signatures; all fell within the acceptable range.

On the Main Deck, Carson sighed in relief. Another blast completed. The asteroid superstructure had creaked and groaned a little. Nothing unusual, but it always put him on edge. One wrong parameter, one hidden crack, and the whole thing could come apart—not withstanding the banding hoops that held everything together.

The monitor displayed the updated countdown to orbital insertion: 92 days, 14 hours, 37 minutes, and 16 seconds. The rotation had wobbled as expected, then stabilized. A small voice in Carson's mind rejoiced. This is what they'd all worked for; the baton had been passed down to him. Generations in the making. He just hoped he could live up to the ghost of their expectations.

After sixteen thousand years of surviving radiation leaks, social upheavals, near-extinction from the Recycler Plague, and the spiritual crisis of the Long Night, Carson knew the hardest test still lay ahead: preparing a civilization of deep-space travelers for the colonization of a new planet—Eridani-B, an Earth-like analogue promising a new dawn. The ship was all they had ever known—for some, all they ever wanted to know.

As if reading his thoughts, Aldo spoke. "One burn closer, Captain." His gaze lingered on the growing disc of 82 Eridani. "One burn closer to the end of everything we've ever known."

— 3 —
THE SHARD

The vacuum drill chattered in Ronny Hayes's gloved hands as he guided it deeper into the asteroid core. Sweat slicked his back beneath the heavy mining suit, the ice-rich vein he'd been following for three shifts promising a good haul. Each cubic meter extracted meant more tokens in his pocket—and more nights at the Helix Lounge with synthesized whiskey and willing company. The thought brought a grin to his weathered face as he leaned into the controls.

The resistance came gradually at first. The drill bit slowed, requiring more pressure, more power. Ronny frowned, adjusting his stance for leverage. "Come on, you bastard," he muttered, increasing the thrust. The bit squealed against something unyielding—not the usual sound of hitting a dense mineral pocket. This was different. Wrong.

"Just needs more juice," he growled, cranking the power to maximum and throwing his substantial weight behind the machine. For a heartbeat, nothing happened. Then chaos erupted— the drill violently kicked back, tearing from his grip like a living thing. It flew across the chamber with impossible force, smashing against the far wall in an explosion of metal fragments and rock dust. The echo of its impact reverberated through the narrow passage.

Ronny staggered back, boots slipping on loose grit. "Son of a bitch!" The curse echoed as he caught himself, chest heaving. Dust clouded his visor; it gradually cleared in the dim light. Then he saw it—a black, crystalline chunk lodged in the exposed rock. Trapezoid-shaped, with surfaces smooth as polished glass, it sat in stark contrast to the rough stone surrounding it.

"Control, Hayes here. Got something weird," he said into his comms unit. "Drill's busted, and there's some kind of … I don't know, crystal or something."

Up in the Mining Control Room, Devory Murchison squinted at Ronny's feed. The video quality from the mining suits was never great, but even through the static and poor lighting, he could see something unusual.

"Crystal? You sure about that?" His tone was dry—Ronny had a reputation for dramatizing routine issues. But the image on the screen nagged at him. The object was too sharp-edged, too geometrically perfect to be a natural formation in the asteroid's core.

"No," Ronny's voice crackled through his comms. "But you might want to see this yourself, Chief."

Devory sighed, pushing himself out of his chair. "Hold tight. I'm coming down."

The cage elevator rattled as it descended into the shaft, the aged mechanism groaning in protest with every meter. Devory had filed three requests for a maintenance overhaul, but with the imminent arrival at 82 Eridani, all resources were focused on the Landing Craft. Mining equipment was low priority—they only needed enough oxygen extracted from the ice to make planetfall, then this hollowed-out rock would be abandoned forever.

His boots crunched on the loose gravel as he stomped up to Ronny. "Show me."

Ronny jerked a thumb toward the drill site. "Right there. Killed my bit."

Devory approached cautiously, a hand drill in his grip. As he extended it toward the black object, something impossible happened—the drill jerked back violently, as if repelled by an invisible force, and lodged in the wall behind them.

His jaw tightened. "No way," he muttered, stepping closer. He grabbed a spare drill bit from his belt pouch and lunged toward the object. The result was the same—the metal was ripped from his grip before it could make contact.

Ronny smirked behind his visor. "Told ya."

Devory stood transfixed, studying the anomaly. It was approximately 30 centimeters long, with that distinctive trapezoid shape and impossibly smooth planes that glinted in the dim light of the mining shaft. This wasn't a crystal or a rock—it was something else entirely.

"Get me a net and some straps," he ordered over the comms. "Now."

Minutes later, Cory Dunbar, second in lead with a mop of red hair that seemed to defy the confines of his helmet, hustled down with the requested gear.

"What's this about, boss?" he asked, eyes darting between Devory and Ronny before settling on the black object embedded in the rock.

"You'll see," Devory said, attention never leaving the anomaly. "Hayes, clear it—picks only."

Ronny grabbed a pick, his suited bulk straining as he hacked at the regolith surrounding the object. The metal tips skittered whenever they came too close to the black surface, as if fighting against the object's push, but he dug in, grunting with effort. Cory joined in, and their suits creaked with exertion as they worked, dirt and rock fragments flying until they had created a rough pit around the object.

"Net it," Devory ordered once they had cleared enough space.

They shoved the net under the Shard, its synthetic strands stretching under tension. They each took a side, entwined their bulky gloved fingers, and yanked it taut.

"Lift," Ronny rumbled.

They heaved—and nothing happened. The object didn't budge. Ronny, a hulk of a man even without the suit, cursed as his muscles bulged with strain. Still nothing. Cory wheezed, his lungs near bursting point.

They redoubled their efforts, working together. This time they managed to roll the object, inch by agonizing inch, into the center of the net. Sweat beaded on Ronny's brow inside his helmet. "Thing's tiny—how's it this heavy?"

Devory crouched, his headlamp gliding over the Shard's trapezoid planes, careful not to touch it. About thirty centimeters long, no fractures, just sharp angles at odds with the rubble around it. Its surface didn't reflect light but swallowed it, then pulsed faintly, as if amplifying it. "Weirdest thing I've ever seen," he said, half to himself. "Doesn't fit any rule I know."

The trip to the surface proved as challenging as the extraction. The Shard—as Devory had taken to calling it—fought them every step of the way. They hauled it to the cage elevator, the net swaying between them, the elevator frame groaning as they loaded it in.

A sudden jolt rocked the cage. The Shard started hovering, it repulsive force buckling the metal floor underneath, the cables overhead buzzing like strung nerves under tension. Cory braced himself against the shaft wall, muttering, "If this thing snaps, it's on you, boss."

Up they went, their breaths tight in their chests, the whine of straining metal filling the air. By some miracle, they reached the surface without disaster, the net swaying between them, the Shard hovering in the net like a droplet of oil floating on water.

The GeoLab was Devory's fortress of science, its aging equipment arranged with military precision despite GenStar's decay. Petrographic microscopes, their lenses scratched from years of analyzing ice veins for oxygen extraction, lined one wall, their calibration charts taped beside them in Devory's neat script. A battered X-ray fluorescence analyzer, its display flickering from uranium ore dust, sat next to it. Nearby, an X-ray diffraction unit, patched with epoxy to seal hairline cracks, hummed faintly, its last scan of uraninite crystals still logged. A mass spectrometer, its vacuum pump wheezing, stood ready to analyze uranium isotopes for reactor fuel, while a jaw crusher's scarred steel teeth loomed over a coring drill press repurposed for ice core extraction. A laminar flow clean bench, its filters yellowed but functional, ensured contamination-free prep for oxygen-critical samples. The magnetic susceptibility instrument twitched erratically, its needle spiking as the Shard's field interfered, defying Devory's scientific control.

The Shard hovered in its net, bending the metal platted deck beneath it as they pulled it to a beam. After slinging straps over, they raised it to a height where its repulsion seemed to fade. "Hold steady," Devory said, watching it settle. "The effect weakens with range." He exhaled sharply, his controlled demeanor cracking. "Unbelievable. Un-fucken-believable."

The team circled the suspended object, wary, treating it like a caged beast. No known mineral matched this specimen—too dense, too perfect in its geometry. It broke physical rules Devory didn't even know existed. A wry grin tugged at his mouth, his pulse quickening despite his professional detachment. What a time to upend everything!

He shook himself, as if breaking a trance. Carson needed to know. This was beyond protocol. He activated his comms. "Devory here. Heads up. We've got something out of spec in the GeoLab. Secured it, but it's cost us oxygen haul hours. Could be trouble—thought you should know." His tone stayed flat, but the oxygen delay stung. With three months to orbital insertion, every hour counted.

A tense silence fell over the lab. Ronny lingered near the door as Cory secured the Shard, his eyes fixed on the hovering anomaly.

"We should've buried it back down there," Ronny muttered. "Feels like we've dug up trouble."

Devory waved off the concern—this was no time for superstition—but the unease lingered like an ion charge. Was this thing dangerous? Were they poking at something best left undisturbed? He pushed the thoughts aside. Rivel would set them straight, he was sure. He could always rely on her. And if not Rivel, then Mia. But he didn't want to think about that right now. The shift wasn't over. The Shard could wait; oxygen quotas couldn't.

– 4 –

COUNCIL MEETING

The Command Center's circular chamber hummed with tension as Captain Frederick Carson studied the holographic display of 82 Eridani. The star's image rotated slowly above the central table, casting amber light across the faces of the twelve people seated around it. Three months until orbital insertion—a blink of an eye after sixteen thousand years of travel—and now this unexpected complication. Sitting immediately to his right was First Officer Howard Norton. Unlike Carson, whose ancestors had been officers since the fourteenth millennia of travel, Norton had risen from the agricultural sector through merit and determination. His uniform bore the subtle modifications of someone not born to command—extra pockets, reinforced elbows, practical adaptations that spoke of his working-class origins. Despite his younger years, his features betrayed the stress of acting as Carson's right-hand man, which left little time for personal life and relaxation.

"Let's begin," Carson said, taking charge. He wore the traditional captain's uniform: a deep blue jumpsuit with silver piping that had evolved over millennia from the original Earth naval designs. The silver insignia at his collar—a stylized asteroid with three concentric rings—gleamed under the artificial lighting. "Okay Murchison, the floor is yours."

Devory straightened in his seat, acutely aware of the eyes upon him. "As reported, during routine oxygen extraction in Shaft 14-B, we encountered an anomalous object embedded in the asteroid core." His voice remained steady, professional, as he eyed those sitting at the table around him. "The object exhibits properties inconsistent with any known material. Most notably, it repels metal with significant force and ... it possesses extraordinary density."

Across the table, Martin Kendricks, Head of Security, glanced briefly at Lora Vasquez, his taciturn younger deputy who was mirroring his rigid posture in an identical uniform—black with red epaulettes that contrasted starkly against the softer blues and grays of the others. There was a flicker of doubt in her eyes that almost went unnoticed.

"Show us," Carson ordered.

Devory nodded to Callum Broadbent, Head of Electronics, who tapped a sequence on his tablet. The holographic star disappeared, replaced by footage from the mining shaft: Ronny Hayes's drill violently kicking back, the black trapezoid gleaming in the exposed rock. The recording continued, showing their attempts to move the object using fiber nets, the strain evident on the miners' faces through their bulbous helmet visors. Vasquez's eyes flicked briefly to Kendricks, a wordless acknowledgment of the anomaly, though she remained as still as a shadow. Anna Teoh, seated near Vandergamma, glanced at her tablet briefly, her brow furrowing as if distracted by her own data.

"We've designated it 'the Shard,'" Devory continued as the footage played. "Current mass calculations indicate it weighs approximately 294 kilograms despite its relatively small size." Reacting to the murmurs that came from the others when they heard this, Devory said, "Yeah, it's denser than a neutron star, if you can believe it. Nearly busted our guts, but we managed to secure it in the GeoLab using non-metallic restraints."

First Officer Howard Norton rubbed his eyes tiredly, then shook himself awake. "And this discovery has delayed oxygen extraction by how many hours?" he asked, his face growing taught with concern.

"Thirteen hours, sir," Devory replied. "We've adjusted shift rotations to compensate, but we're still projecting a net loss of approximately 4.3 percent in this extraction cycle."

Norton shook his head. "4.3 percent?" he grouched, "and this is with adjusted shifts?"

"It's the best we can do at the moment," Devory said, swallowing hard.

Norton looked at Carson. Carson sighed heavily. "Obviously, we're going to have to make up the shortfall over the coming
shifts. What's your contingency for that," he asked Devory.

Devory wrung his hands. "We'll have to double some shift,
extend overtime."

Hearing this, Mia couldn't help interjecting. "If I may?" she
said, speaking up with a hint of tension in her voice. "Doubling
shifts would strain the mining team significantly. They're already
working at capacity. I'd recommend instead—"

"Your recommendations are noted, Dr. Hawthorn," Carson
interrupted, "but this isn't open for debate. The mission timeline
takes precedence."

Mia fell silent, though her eyes met Devory's briefly across the
table, a silent communication born of their personal connection.
She noted the slight tension in his jaw, the way his fingers had
stopped their tapping and now pressed flat against the table's surface. It was evident to her finely trained eye that he was warring
internally over his professional duty and his scientifically aroused
curiosity. For the first time as long as she could remember, the
Shard represented an opportunity to break out of the stale rut
they had all sunken into. But it seemed to have come at a cost she
could not yet fully measure.

"With respect, Captain," Rivel Torvax said, her voice carrying
the slight accent of the Third Ring where engineering families
had maintained their own dialect for generations. Her dark hair

was pulled back in a practical braid that accentuated her ebony features. But this was not her most outstanding feature. It was the black eye patch that covered her left eye, lost in the same accident that killed Hawthorn's father, Eric, giving her a fearsome edge she wielded effortlessly. "We may be the first humans in history to discover an object like this. That alone should give us pause for thought. For the sake of science."

"I'm aware of your scientific interest, Chief Engineer," Carson replied, his tone tight and cool. "But our mission parameters are clear. Oxygen extraction takes priority over everything except critical ship functions."

Rivel exchanged glances with Callum Broadbent, his usually jolly expression was completely absent, his investment in the Shard as palpable as hers. Unlike Rivel, who had inherited her position through her family's long engineering lineage, Callum had earned his role through sheer technical brilliance, overcoming the social barriers that typically limited advancement from the Outer Ring. The contrast in their backgrounds was evident in their bearing: Rivel confident in her authority, Callum persuasive through diplomacy.

"Captain," Callum ventured, his voice softer than Rivel's but no less insistent, "the Shard's metal-repelling properties could have significant implications for our electrical systems. If we could understand the mechanism, it might revolutionize our power distribution efficiency."

"Or pose a threat to ship systems," countered Martin Kendricks, his voice cutting through like a blade. Vasquez turned to him ever so slightly as he spoke, as a sign of dutiful respect. "Unknown materials represent unknown risks," Kendricks said. "My recommendation is we eject it immediately."

"Eject it?" Rivel's voice rose in disbelief. "That would be an unconscionable waste of—"

"Of what?" Kendricks interrupted. "A potential hazard three months before our arrival? The timing is suspicious at best."

"Suspicious?" Dr. Lilien Chen raised an eyebrow, her medical whites faded with age. As Chief Medical Officer, she held equal rank to Norton but approached problems with clinical precision rather than practical compromise. "Based on what Devory has told us, it's a mineral formation, Kendricks, not a saboteur."

"We don't know what it is," Kendricks insisted, his eyes narrowing as Vasquez's mirrored his vigilance. "That's precisely my point."

Prya Chandra, who had been listening silently, finally raised a hand. The simple gesture commanded immediate attention, a testament to the spiritual authority she wielded. Her robes, deep purple with intricate geometric patterns representing cosmic cycles, rustled softly as she leaned forward.

"Perhaps," she said, her dark eyes taking in each speaker, "we should consider what the Continuance teaches us about unexpected discoveries." She looked directly at Carson. "The Eighth Reflection warns of 'signs that appear at journey's end'—markers of transition that test our readiness for arrival. I think there is something significant in these words in lieu of what has been discovered." Dr. Chen gave a subtle nod, acknowledging Prya's point while maintaining professional detachment.

Carson's expression remained neutral, though those who knew him well could detect the slight softening around his eyes. His relationship with Prya was complex—publicly professional, privately more dependent than he would ever admit. The balance between practical and spiritual authority had sustained the GenStar for millennia, and despite his focus on mission parameters, Carson understood the crew's need for meaning beyond technical objectives.

"The Continuance offers wisdom," he acknowledged with a slight nod to Prya, "but our immediate concerns are practical. Belief alone won't fill our oxygen tanks."

"True," came the voice of the Grand Archivist. The room fell silent—the Archivist rarely contributed directly to these discussions. Heads turned his way, curious to see what this break in protocol would offer. "Based on a detailed index search I just ran," Tiber Solis said, his voice characteristically deep and sonorous, "there's nothing like it in all the archives. This includes the transit logs, mineral surveys, even mythic references—nothing parallels this object's properties." He tapped his data tablet, to confirm what he just said. "This alone makes it significant enough to warrant careful consideration."

Carson regarded Tiber thoughtfully. The Archivist's intervention carried weight precisely because of its rarity. "Your point is noted, Archivist Solis. There is no doubt it requires careful consideration. The question is when."

Tiber inclined his head in acknowledgment, returning to his tablet. He had made his point without directly challenging authority—a skill refined through generations of Archivists who had learned to influence without appearing to do so.

Carson rubbed his chin in thought. Tiber's words seemed to have thrown him off balance. He looked at the Chief Navigator. "Vandergamma, what's your assessment? Can we afford to divert man-hours to this thing?"

Aldo Vandergamma, the Chief Navigator whose family had guided the GenStar for nine generations, cleared his throat. His uniform bore the distinctive silver stars of the Navigation Corps, matched by his thinning silver hair, which had turned that color abruptly after the Pneumatic Blowout nineteen years prior.

"Our approach vector remains stable," he reported, grateful for the shift to familiar territory. "The last deceleration burn achieved 99.7 percent of projected delta-v. We're on schedule for orbital insertion in 92 days, 11 hours. If the double shifts in the mining sector recover our lost quota, then I see no reason for us to over-react. I do, however, recommend that we refocus our efforts on the Landing Craft. Any further distractions will only increase the likelihood of an error in our preparations, and we all know we can't afford that."

"Well spoken," Carson said, relieved. Vandergamma's authority had ascended above all others in these last months, as it should, trumping even the Grand Archivist. But his relief was short-lived.

Vandergamma added, "And speaking of error," he said quietly, "Anna has been running over some trajectory calculations and some things don't appear to be adding up."

Carson raised an alarmed eyebrow. "But didn't you just say our approach vector remains stable?"

"It is. However—" He looked at Anna.

Anna Teoh imperceptibly straightened. "All standard models show successful insertion within acceptable parameters," she stated, her voice crisp. "However ..." She hesitated, a behavior Mia noted was uncharacteristic for someone who usually was confident of her calculations.

"However?" Carson prompted.

"I've been exploring some alternative mathematical frameworks," Anna continued, choosing her words carefully. "There are patterns in our approach that conventional models don't fully capture, harmonics ... nothing concerning," she added quickly, noting Kendricks' narrowing eyes and Vasquez's subtle tilt of the head, "I think it's just a case of throwing more computing power at the problem."

Mia Hawthorn, who had been observing the dynamics of the meeting with professional interest, noted the subtle shift in Anna's speech patterns—a slight increase in abstraction that wouldn't be apparent to anyone not specifically looking for it. As the ship's Cognitive Neuroscientist, Mia's role in these meetings was ostensibly to advise on crew psychological readiness, but her training made her acutely aware of changes in cognitive patterns.

"Could these harmonics be related to gravitational anomalies?" Mia asked, directing her question to both Anna and Philip Polsen, the Cryogenics Chief whose private expertise in quantum effects made him the closest thing they had to a theoretical physicist.

Polsen adjusted his glasses. One of the few who had opted against laser surgery to correct their eyesight. His position as head of Cryogenics gave him authority not only over the fueling of the Landing Craft, but over the frozen embryos and DNA samples that represented humanity's genetic diversity, a responsibility that had made him methodical to the point of obsession.

"Unlikely," he replied after careful consideration. "Gravitational anomalies would affect our instrumentation uniformly. What Anna describes sounds more like quantum interference patterns." He turned to Anna. "Are you accounting for non-local effects in your models?"

Before Anna could respond, Vasily Petrov, the Propulsion Chief responsible for the nuclear deceleration system, cut in with characteristic impatience. "This is fascinating academic discussion," he said, his accent thickened by irritation, "but we're ignoring the practical problem. The oxygen extraction delay affects my department directly. The Landing Craft's fuel requirements are non-negotiable."

"Agreed," Carson reaffirmed. "Which is why I'm making an executive decision."

The table looked at him, waited for the hammer to come down.

"From this point on, no one is to spend any more time on the object. Do I make myself clear?" He eyed each person in turn, then softened a little, allowing a conciliatory tone to wash over him. "I understand we are dealing with the unknown here. But my job is to ensure that we first deal with what we *do* know. And that is preparing the Landing Craft for orbital insertion. Once we achieve that, we can revisit this discussion. At that time, I will be open to all suggestions, so long as it doesn't jeopardize the mission." He looked directly at Rivel, hoping to secure her agreement. "For now, just remember our motto: We carry the weight of our ancestor's sacrifice."

The invocation of their ancestors—the ultimate appeal to authority aboard the GenStar—settled over the room like an invisible shroud. Even Rivel, with all her intellectual independence, couldn't argue against the generations of sacrifice that had brought them to this point.

Carson looked around the table one last time. Satisfied that he had made his point. "Then this meeting is adjourned," he declared, rising from his seat. The others followed suit, the hierarchy of departure as formalized as everything else aboard the GenStar: Carson first, followed by Norton, then the department heads in order of their sectors' proximity to the ship's core systems.

As they filed out, Mia lingered, watching the interactions with professional interest. Rivel and Callum huddled together, their whispered conversation intense despite the captain's clear orders.

Prya approached Carson, her hand lightly touching his arm in a gesture that appeared casual but carried the weight of their complex relationship. Kendricks and Vasquez stood apart, their matched uniforms a silent declaration of vigilance, his suspicion mirrored in her steady, unblinking gaze. Anna remained at the table, staring at her calculations with an abstracted expression that hadn't been present in previous meetings.

Devory approached Mia as the room emptied, his voice low. "Dinner later? Common Dome?"

Mia nodded, noting the unusual intensity in his eyes. "Everything all right?"

"Fine," he replied too quickly. "Just want to discuss something ... interesting."

As they left the Command Center, Tiber Solis remained behind, making final notes of the meeting. He glanced up at the holographic display, now returned to its default view of 82 Eridani growing incrementally larger with each passing day. Sixteen thousand years of human history had led to this moment—generations living and dying aboard the GenStar, all to reach a star that now hung before them, tantalizingly close.

And three months before arrival, they had discovered something that defied sixteen thousand years of accumulated knowledge.

Tiber made a private notation in a personal file, not part of the official record, but a continuation of the personal observations maintained by Archivists since the beginning of the mission and which he kept on a data stick attached to a lanyard around his neck. His entry was simple but would prove prophetic in the days to come:

Today marks a divergence from all historical patterns. The Shard represents an unprecedented variable in our carefully plotted course. I fear we are no longer following the path laid out by our ancestors, but entering uncharted territory with no historical guidance to light our way.

— 5 —
THE COMMON DOME

T he Common Dome opened beneath them like a vast, inverted bowl sunk into the floor. Its five-meter-thick reinforced glass revealed the black void beyond, a startling window into the cosmos. A waist-high railing encircled the perimeter, a practical reminder of the drop. Mia paused at the entrance, momentarily transfixed by the sight of 82 Eridani—a brilliant disc of light suspended in the darkness like a luminous jewel floating in black ink. Layers of water sandwiched between the glass panes absorbed cosmic radiation while keeping the view pristine, creating the uncanny illusion that nothing separated the diners from the vacuum of space below.

The Dome had been designed generations ago as a psychological necessity—a place where Arkonauts could gather beneath something resembling a sky. Over centuries, it had evolved into the social heart of the GenStar, where hierarchies relaxed slightly and the rigid compartmentalization of ship life gave way to something almost resembling the communities their ancestors had left behind.

"Hungry?" Devory asked, his hand lightly touching the small of Mia's back.

Mia hadn't noticed him come up behind her; she had been so absorbed by the sight of Eridani. She pretended not to be surprised and smiled, allowing herself to be guided toward the food dispensary. The evening crowd had already formed into its usual patterns—engineers clustered near the center tables, agricultural workers claiming the spaces nearest the hydroponics display, navigation personnel predictably positioned with clear sightliness to the stars. The social geography of the Common Dome was as fixed as the physical geography of the GenStar itself.

They collected their trays and joined the line, the sounds of the kitchen mixing with the continuous chatter of voices. Ahead, a group of maintenance workers debated the merits of the latest recycler modifications, their hands gesturing emphatically as they shuffled forward.

"Boiled grains with legumes," Mia requested when they reached the front. The server—a young woman with the distinctive pale complexion of those born in the agricultural ring—nodded and ladled the mixture onto her plate. "And coriander with yogurt, please."

"Synthetic or bacterial culture?" the server asked, her tone suggesting the question was purely procedural.

"Synthetic," Mia replied. The bacterial culture yogurt was technically more nutritious, but the synthetic version reminded her of her childhood—a small comfort in days that increasingly offered few.

Devory opted for a lentil burger with beetroot and kale, his choice as predictable as everything else about him had been until recently. Mia studied his profile as they moved along the line, noting the subtle changes in his demeanor—the slight acceleration in his movements, the unusual brightness in his eyes. Nothing dramatic enough to raise alarm, but noticeable to someone who had shared his bed for the past three years.

They found a table slightly removed from the main seating area, positioned in the "southern" quadrant of the Dome where the ambient lighting was dimmer. Privacy was a relative concept aboard the GenStar, but this was as close as they could get to a private conversation in a public space.

"So," Mia said as they settled into their seats, "are you going to tell me what's got you so energized, or do I have to guess?"

Devory glanced around before leaning forward, his voice dropping. "It's the Shard," he said, unable to keep the excitement from his tone. "Mia, it's unlike anything we've ever encountered. The properties it exhibits—they shouldn't be possible."

Mia took a measured bite of her food, chewing slowly as she

observed him. "The Captain seemed less than enthusiastic about your discovery."

"Carson's focused on the mission timeline," Devory replied, dismissing the concern with a wave of his hand. "He can't see beyond orbital insertion. But this ..." He paused, searching for words. "This could change everything we understand about physics."

"That's quite a claim," Mia said carefully. "Based on what evidence?"

Devory's eyes lit up at the question—more like a little boy than a grown man. "The metal repulsion is just the beginning," he said, leaning closer. "We've been running tests—"

"Tests? You heard what Carson said," Mia interjected, keeping her voice low.

"Basic safety protocols," Devory countered smoothly. "Nothing that interferes with extraction duties."

Mia raised an eyebrow but didn't press the point. "And what have these 'safety protocols' revealed?"

"It absorbs all radiation without reflection or emission," Devory said, his food forgotten as he raced on. "We've tried everything—visible light, infrared, ultraviolet, even low-level gamma. It just ... takes it all in. No heat signature, no reflection, nothing."

"That violates basic thermodynamic principles," Mia noted, her scientific curiosity reluctantly piqued despite her concerns.

"Exactly!" Devory's voice rose slightly before he caught himself, glancing around before continuing more quietly. "And that's not all. Its temperature remains constant regardless of external conditions. We tried cooling a section with liquid nitrogen—nothing. Heated another section with a plasma torch—nothing. It's as if it exists in its own thermodynamic bubble."

Mia frowned. "That's not possible."

"Neither is negative matter," Devory replied, "which is what Rivel thinks. And you know she's not dumb."

"Rivel?" Mia said, almost choking on the word, a flicker of something—not quite jealousy, but adjacent to it—passing through her. "You've been spending a lot of time with her."

Devory blinked, momentarily derailed. "She's the Chief Engineer," he said, as if that explained everything. "Her expertise in materials science is crucial for understanding the Shard's properties."

"Of course," Mia said, returning her attention to her food. "Go on."

If Devory noticed the slight shift in her tone, he gave no indication. "Rivel's theory is the only one that makes sense—theoretical particles with negative mass. It would explain the repulsion effect and the thermodynamic anomalies."

"It could," Mia agreed, "but why here, why now?"

Devory nodded as if anticipating the question. "The only explanation is it must be a remnant of a primordial black hole. Hawking theorized they formed before the first stars, before the universe had stabilized, when all kinds of exotic materials were still in play."

"And it just happened to lodge itself in our asteroid?"

"Yes!" Devory effused, his eyes bright with the fervor of discovery. "Think about it, Mia. If we're right, this is the most significant scientific discovery since relativity. Hell, maybe since the first stone tools."

Mia studied him, noting the subtle changes in his speech patterns—the increased rate, the hyperbolic comparisons, the slight tremor in his hands as he finally remembered to take a bite of his food. Nothing dramatic, nothing that would trigger alarm in someone who didn't know him intimately. But to her trained eye, the shifts were unmistakable.

"Have you considered the possibility that the Shard might be affecting you?" she asked carefully.

Devory's expression clouded. "What do you mean?"

"I mean," Mia said, choosing her words with precision, "that an object with properties as unusual as you're describing might have effects beyond the purely physical. Cognitive effects, perhaps."

"You think it's altering my mind?" Devory's became clipped, defensive. "Based on what evidence?"

"Your enthusiasm is ... atypical," Mia replied, maintaining her clinical detachment despite the personal nature of the observation. "You've always been methodical, cautious. This excitement over an unverified hypothesis isn't your usual approach."

Devory's jaw tightened. "So scientific curiosity is now a symptom?"

"I didn't say that," Mia countered gently. "But as a neuroscientist, I'm trained to notice cognitive shifts. And there's something different about you since this discovery."

For a moment, tension hung between them—the professional observation colliding with personal intimacy. Then Devory's expression softened slightly. "Maybe I am a bit more excited than usual," he conceded. "But wouldn't you be? This could be the discovery of a lifetime."

Mia nodded, acknowledging the point while still harboring her concerns. "Which is why it should be studied properly," she said. "Including potential cognitive effects."

Devory's eyes lit up again. "Are you offering to help?"

The question caught her off guard. She had intended to urge caution, not volunteer participation. But now that he'd suggested it ... "I could monitor neural responses," she said slowly, her own scientific curiosity beginning to stir despite her reservations. "If the Shard does have cognitive effects, they should be measurable."

"Carson would never approve," Devory said, though his tone suggested this was hardly an insurmountable obstacle.

"No," Mia agreed. "He wouldn't."

They ate in silence for a moment, the unspoken question hanging between them. Around them, the normal rhythms of the Common Dome continued: laughter from a table of maintenance workers; the soft chime announcing the evening nutrient supplements; the gentle hum of the air recyclers maintaining the perfect balance of oxygen and carbon dioxide.

"Rivel thinks we might be able to break off a fragment," Devory said finally, his voice low. "She's designing a modified hydraulic hammer from a non-metallic alloy. If it works ..."

"We'd have a piece that could be studied outside the GeoLab," Mia finished the thought.

Devory nodded. "Away from Kendricks' surveillance."

Mia should have shut down the idea immediately. The rational, responsible course of action was clear. Report the unauthorized testing, insist on following Carson's orders, prioritize the mission timeline, observe sixteen thousand years of ancestry with dogmatic inflexibility. Instead, she heard herself asking: "But what if ... Carson does find out?"

"He won't." Devory caught her hand, his grip warm. "We'll keep the fragment in your lab. No one monitors your neural research."

Because no one cares until it's too late.

The thought slithered through her before she could stop it. She shoved it aside.

"One condition," she said.

"Name it."

"I scan you first. Full neural workup."

Devory stiffened. "Why?"

"Baseline data," she lied. "If the Shard affects cognition, I need to know before we handle a piece of it."

For a heartbeat, she thought he'd refuse. Then he relaxed. "Deal."

Below them, 82 Eridani shone through the Dome, its light casting elliptical shadows on the ceiling. Three months until orbital insertion, the culmination of sixteen thousand years of human perseverance. And yet here they were, planning to study an anomalous object against direct orders, potentially risking the mission timeline for scientific curiosity.

Mia wondered what their ancestors would think of their choices. Would they understand the drive to know, to discover, to push beyond established boundaries? Or would they see only the risk to the mission they had sacrificed everything to maintain?

As she looked at Devory, seeing his eyes bright with excitement, his hands gesturing as he outlined their approach, she felt a flicker of unease. Something was changing in him, something subtle but undeniable. And instead of stopping it, she was about to become complicit in its progression.

"Tomorrow morning," she said, making her decision. "Come to my lab before your shift. We'll establish the baseline, then proceed from there."

Devory smiled, squeezing her hand. "You won't regret this, Mia. We're on the verge of something extraordinary—I can feel it."

So can I, she thought, but the feeling taking root in her chest wasn't excitement. It was something darker, more primal. An icy warning that they were crossing into the unknown. But the scientist in her—the part that had driven her to study the human mind in all its complexity—couldn't turn away from the mystery. If the Shard was affecting Devory's cognition, she needed to understand how and why. And if it wasn't ... then what they had discovered truly was extraordinary.

Either way, as of tomorrow, she would be part of whatever the Shard had set in motion. For better or worse.

Around them, the Common Dome bustled with the ordinary rhythms of ship life—Arkonauts eating, talking, laughing, unaware that in a quiet corner, two of their number had just taken the first step down a path that would alter the course of their sixteen-thousand-year journey in ways none of them could imagine.

— 6 —
COGNITIVE NEURO LAB

Mia glanced at the time display: 04:37. Early enough that the corridors were nearly empty, late enough that her presence wouldn't trigger automatic notifications. The perfect window for activities that fell outside official protocols.

Her lab was small but efficient, a sanctuary of order amid the controlled chaos of the GenStar. Unlike the communal workspaces that dominated most of the ship, she had carved out this small domain for herself, a testament both to her status as the ship's neuroscientist and to her skill in navigating its byzantine bureaucracy. The room was barely six meters square, yet over time she had managed to squeeze in everything that was important. Neural imaging equipment lined one wall, while cultivation chambers for her organoids occupied another. Cupboards brimmed with glassware, reagents, and other consumables. A wet area with a sink, a small cryostat, and a microscope stood nearby, the latter flanked by computer monitors where she could configure multiple holographic displays for analysis.

The door slid open with a soft hiss. Devory entered, his movements quick and furtive. It was so unlike his steady gait, it stirred a fresh ache of worry in Mia. Dark rings hollowed his eyes, and a fine sheen of sweat glistened on his forehead despite the lab's carefully regulated temperature.

"You look terrible," she said by way of greeting.

Devory managed a tired smile. "Good morning to you too."

"When did you last sleep?"

"Define 'sleep,'" he countered, dropping into the chair opposite her workstation. He raked a hand through his tangled hair. "We've been working on this for three shifts straight."

"We?"

"Rivel and I." He leaned forward, lowering his voice despite the privacy of the lab. "It wasn't easy, Mia. That thing," he grimaced, "it fought us every step of the way."

Mia plugged in her data stick and activated the room's privacy veil, a precaution she'd installed years ago when her research into cognitive patterns had veered into politically sensitive territory. The ambient lighting shifted slightly, indicating that both audio and visual monitoring had been disabled. The feed to Martin Kendricks' Control Room would go dark. Not that he'd notice at this hour, but later on he would ask why. She'd have to think up an answer ahead of time, but right now, that could wait.

"Tell me," she said, studying him with both personal concern and professional interest.

Devory's eyes brightened, that now-familiar excitement cutting through his exhaustion. "Rivel's hydraulic hammer took some punishment." His hands moved as he spoke, sketching the device in the air. "The Shard resisted. It's like ... it's like it knew what we were trying to do."

Mia frowned. "That's an unusual choice of words. 'Knew'?"

Devory waved away the concern. "Figure of speech. The point is, conventional approaches failed completely. The hammer just bounced off, no matter how much force we applied."

"So how did you succeed?"

"Resonance." A smile spread across his face. "Rivel had the insight. Instead of brute force, we tried a harmonic approach. We calibrated the hammer to deliver pulses at specific frequencies. We cycled through the spectrum until we found one that the Shard responded to. I'm not saying ... how should I say?"

"What frequency?"

"40 Hertz." Devory's expression turned thoughtful. "Interesting number. Not what I would have predicted."

Mia made a mental note of the figure. 40 Hz fell within the gamma wave range of human neural activity—the frequency associated with consciousness and attention. A coincidence, surely, but one worth investigating.

"Once we found the resonant frequency," Devory continued, "we were able to create a stress fracture along one of the geometric planes. Even then, it took hours of sustained application before a fragment finally separated." He reached into his pocket and withdrew a small glass vial, sealed with a polymer cap. "And here it is. You're the lucky winner."

Gazing at it, Mia observed a small fragment—no more than two centimeters in size—cut a stark silhouette against the curved glass. Its surface was as smooth and geometrically perfect as the parent object, the edges forming precise angles that seemed to defy any natural formation process. Despite its size, its allure gripped her like nothing she'd known: no wonder Devory was so enthralled by it. Still, she checked herself, an instinctive response born out of abundant caution.

"You're sure it's safe in there?" Mia asked, making no move to take the vial.

"The glass is non-metallic, and we've tested the fragment's repulsion field. It's proportional to mass. This piece only affects metal within a few millimeters of contact." He held out the vial. "Perfectly safe for your experiments."

Mia hesitated, then took the container, careful not to touch Devory's fingers in the process. The vial felt much heavier than it should. She felt like her brain was playing a trick on her: nothing this size could have that much mass.

"Boy, it's heavy," she exclaimed, holding it up to her eyes in wonderment.

"That's nothing. You should have seen us lug the parent body."

"I can imagine."

Mia continued to stare at the specimen. "It's going to be interesting to see what comes out of this."

"Like whether I'm losing my mind?" Devory's tone was light, but his eyes held a challenge.

"Like whether the Shard has properties beyond the purely physical," Mia corrected him gently. "You agreed to the neural scan, remember?"

Devory sighed but nodded. "A deal's a deal." He glanced at the vial. "But after you've run your tests, I want to know everything you find. No holding back, even if it's ... concerning."

"Of course." Mia carefully put the vial in a rack on her workbench. She gestured toward the neural imaging station at the back of the room. "It won't take long. Standard cognitive mapping, nothing invasive."

As Devory settled into the scanning chair, Mia prepared the equipment with practiced efficiency. The neural cap, a web of quantum sensors tuned to catch the faint whispers of brain activity, hugged his skull like a second skin. She calibrated the sensitivity, adjusted for ambient electromagnetic noise, and initiated the scanning sequence.

"Try to relax," she instructed, watching the initial readings stabilize. "Think neutral thoughts."

Devory closed his eyes, his breathing slowing as he complied. The holographic display above the workstation bloomed with color, a three-dimensional representation of his neural activity. Mia studied the patterns with a frown. The prefrontal cortex showed heightened activity, consistent with his apparent excitement and focus on the Shard. But there was something else. Subtle fluctuations in the temporal lobe that didn't match standard cognitive patterns.

"Are you experiencing any unusual sensory perceptions?" she asked, keeping her tone casual. "Visual anomalies, phantom sounds, that sort of thing?"

"No," Devory replied without opening his eyes. "Why?"

"Just covering all bases." She adjusted the scan parameters, focusing on the temporal anomalies. "A few more minutes."

The deeper scan revealed more unusual patterns—coherence between regions that typically operated independently, synchronization at frequencies that didn't align with standard neural oscillations. Nothing dramatic enough to indicate pathology, but definitely atypical. The power spectrum was elevated at 40 Hz. Like his brain had been slightly amplified.

The same frequency that had allowed them to fracture the Shard.

Mia saved the scan data and removed the neural cap, her expression carefully neutral. "All done. Thank you for your cooperation."

"And?" Devory prompted, sitting up. "Am I still sane?"

"Your neural patterns show some unusual activity," she admitted, seeing no point in complete deception. "Nothing concerning, but worth monitoring. I'd like to do follow-up scans over the next few days."

Devory studied her face, clearly trying to determine if she was withholding information. "But you'll proceed with the experiments?"

"Yes," Mia said, making her decision. "I'll proceed."

Relief washed over his features. He stood, swaying slightly with exhaustion. "I should get back before anyone notices I'm missing. Extraction shift starts in an hour."

"You should rest first," Mia suggested, genuine concern coloring her voice.

"No time." He gestured toward the vial. "But this is worth it, Mia. You'll see."

As he turned to leave, she caught his arm. "Devory. Be careful. If Kendricks suspects anything ..."

"He won't." Devory's confidence seemed absolute. "Rivel's covered our tracks. The fragment won't be missed—we took it from an area that's already been documented and sealed."

Mia nodded, not entirely reassured. "I'll contact you when I have results."

"I'll come by after my shift." He hesitated, then leaned in to kiss her—a brief, almost perfunctory gesture that lacked their usual warmth. "Looking forward to it."

Mia watched him leave. She kept on watching as the door closed.

Slowly, very slowly, she turned to the vial. The fragment

seemed to devour the light, its color was so intensely black, yet its outer surface still gleamed like an exquisitely polished mirror. It was just her and the Shard now. She thought of Devory's neural scan, the unusual patterns and the 40 Hz rhythm that matched the Shard's resonant frequency. Coincidence? Or something more troubling?

She needed to find out.

She moved to the cultivation chamber where her brain organoids were maintained, and snapped on a pair of latex gloves. It was time to get to work. The organoids, grown from pluripotent cells then differentiated, allowed her to study the effects of "drift," her primary research question: what is the best way to measure what happens to people's brains when they have been stuck in a can for hundreds of generations. It all came down to the synaptic connections, the way neurons "talked" to each other. With these organoids, she could literally watch it in real time.

Mia selected one of the more mature specimens, a sixty-day culture that had developed robust neural connections and spontaneous electrical activity. With careful precision, she transferred it to the multi-electrode array chip that would allow her to monitor its electrical patterns. The organoid, floating in nutrient medium, pulsed with faint bioluminescence, a genetic modification she had introduced to visualize calcium waves without external markers.

She activated the two-photon microscope, calibrating it to trigger and track calcium waves through the organoid's structure. The baseline measurements appeared on her display: normal signal propagation, with neural impulses decaying over distances of 50-100 nanometers. The standard conduction length for microtubules in neural tissue.

Next, she prepared the exposure chamber. A diamond-coated microfluidic device designed to shield the organoid from electromagnetic interference while allowing controlled exposure to the test subject. She positioned it carefully, then hesitated before placing the Shard fragment.

What she was about to do violated multiple protocols. Carson had explicitly forbidden further testing of the Shard. Kendricks would consider this a security breach. If discovered, she would face severe consequences, possibly even charges of endangering the mission.

But the scientist in her couldn't turn away. If the Shard was affecting Devory's cognition, she needed to understand how and why. And if it wasn't ... then at the every least they had discovered something extraordinary.

She placed the vial in the chamber, positioning it five centimeters from the organoid so that it was close enough for interaction, but far enough to prevent direct contact. The boron nitride walls would block any conventional electromagnetic effects, isolating the experiment to whatever unusual properties the Shard might possess.

"Beginning exposure sequence," she murmured to herself, initiating the protocol. "Thirty-second intervals, monitoring for resonance at 40 Hz."

The first exposure period passed without notable effect. The organoid's neural activity continued its normal patterns, calcium waves propagating and decaying as expected. Mia frowned, adjusting the monitoring parameters. Perhaps the fragment was too small to produce measurable effects, or perhaps the boron nitride shielding was blocking whatever influence the Shard exerted.

During the third exposure interval, something changed, a subtle shift in the organoid's electrical patterns, nothing dramatic, but a definite alteration in the rhythm of neural firing. Her pulse quickened, not just from fear, but from the thrill of the unknown. She edged nearer, tracking the shift as it rippled through the synaptic connections like a breath on still water.

"Anomaly detected," she noted, recording the timestamp. "Synchronization increasing across neural network."

By the fifth exposure, the changes were unmistakable. The organoid's neural activity had shifted to a more coherent pattern, with groups of neurons firing in synchronized bursts. More significantly, the calcium waves were propagating further than should be physically possible, crossing multiple cell layers before decaying.

Mia triggered a test pulse at the edge of the organoid, watching in astonishment as the signal traveled through the neural network with minimal decay. The microtubule conduction length had increased dramatically—from the baseline of 50-100 nanometers to approximately 500 nanometers. A five-fold increase that defied conventional understanding of neural physics.

"Impossible," she whispered, running the measurement again to confirm. The result was the same.

She switched to the quantum coherence simulation, feeding the organoid's microtubule configuration into the molecular dynamics model. Under normal conditions, quantum effects in biological systems were limited by thermal decoherence—the noise of molecular motion at body temperature disrupted quantum states within nanoseconds, restricting them to subatomic scales.

The simulation results appeared on her display, and Mia stared in disbelief. The model predicted coherence lengths of approximately 200 nanometers—ten times the theoretical maximum at biological temperatures. According to every established principle of quantum biology, this should be impossible.

Yet the evidence was before her. The Shard fragment was somehow extending quantum coherence in biological systems, allowing microtubules to maintain quantum states across distances that bridged multiple neurons. The implications were staggering—and disturbing.

Mia ran the control checks, exposing a second organoid to the same conditions but without the Shard fragment present. No changes in conduction length or coherence patterns. She then shielded the fragment with a thin metal plate. Immediately, the effects in the first organoid began to fade, returning to baseline within minutes.

The Shard was definitely the source of the effect. But how? And why?

She was preparing a more detailed analysis when she noticed something unexpected in the neural firing patterns. The exposed organoid was generating sequences that didn't match its previous activity. It displayed complex patterns that seemed almost ... familiar.

With growing unease, Mia compared the patterns to her database of neural recordings. The match wasn't perfect, but it was statistically significant: the organoid was firing in sequences that resembled patterns she had just recorded from Devory's brain—a faint trace of his excitement, or perhaps his fear, flickering in the data.

"Memory echoes," she murmured, the implications sending a chill through her. Somehow, the Shard was facilitating the transfer of neural patterns between physically separate systems. It didn't make sense. It was a form of information transfer that ought to be fundamentally impossible.

As she watched, the microtubules in the exposed organoid began to reconfigure, then relax, creating a subtle pulsation of the neuron cytoarchitecture. She had never seen this before either.

Fearing the moment might pass, she swiftly guided her hands over the controls, documenting every aspect of the phenomenon. The Shard wasn't just affecting neural tissue, it was restructuring it, imposing patterns that served some unknown purpose. And if the fragment could do this to an isolated organoid, what might the full Shard be doing to those who worked with it? To Devory?

She thought of his neural scan, the unusual coherence patterns, the persistent 40 Hz rhythm. Not coincidence after all, but evidence of influence—subtle but real. The Shard wasn't just a passive object of study; it was actively interacting with human neural systems in ways they couldn't detect without specialized equipment.

A new and disturbing possibility formed in her mind: What if the Shard wasn't just studying them? What if it was using their neuronal microtubules as some kind of antenna or interface?

Mia glanced at the time display: 07:22. Her official shift would begin soon, and others would be arriving in the adjacent labs. She needed to secure the evidence and the fragment before anyone discovered what she was doing.

With careful precision, she documented her findings, encrypting the data with her personal codes. She returned the fragment to its vial, sealed it, and locked it in her private storage compartment—a small safe embedded in the lab floor that didn't appear on any official schematics.

The exposed organoid presented a different problem. Protocol dictated that experimental samples be preserved for verification, but she couldn't risk anyone else examining it. After a moment's hesitation, she initiated the emergency sterilization sequence, watching as the organoid and its altered neural structures were reduced to their component molecules.

No evidence. No proof of what she had witnessed. Just her encrypted data and the fragment itself.

As she completed the lab sterilization, Mia's mind raced with implications. The Shard's ability to extend quantum coherence in biological systems and facilitate pattern transfer between neural networks represented a fundamental breakthrough—or threat—that went beyond anything in human experience. This wasn't just an unusual material; it was something that could potentially alter the very nature of human cognition.

And Carson wanted to ignore it until after orbital insertion.

Mia saw Carson's reasoning. The mission timeline took precedence after sixteen thousand years of travel. But what if the Shard represented a variable that could affect that mission in ways they couldn't predict? What if its influence was already spreading through the ship's command structure via those who had contact with it?

She thought of Devory, his increasing obsession, the subtle changes in his behavior and neural patterns, all stemming from that strange object unearthed during the last planetary survey. Was he still making decisions based on his own judgment, or was he being influenced in ways too subtle to detect?

The door chime interrupted her thoughts. It was Dr. Lilien Chen coming to say good morning and officially start the shift. She entered the lab with the brisk efficiency of someone who'd spent thirty years rationing both time and trust. The silver streaks in her hair might have softened another woman's appearance, but on Chen they only accentuated the sharp geometry of her cheekbones. Just visible above her collar lay the ghost of a tattoo—three faded characters that were rumored to read: tiān nuò—"Heaven's Promise." The youthful idealism had long since leached from both the ink and the woman, leaving only the pragmatic scientist who now glanced at Mia's gloved hands and simply said, "Oh," without hiding her surprise. Mia peeled the gloves off while turning away, to hide her self-conscious reaction. "I was just doing some cleaning," she said. "In preparation for today's scans."

Dr Chen didn't seem to think much of it. She said, "Sure. Meet me in my office as soon as you're finished."

Mia tossed the gloves in the recycle bin. Every item, including surgical gloves had to be recycled to conserve materials and energy. "I won't be long," she said.

Chen closed the door on her way out.

Mia took a deep breath, composing herself. She would continue her research, carefully and secretly. She would monitor Devory for further changes. And she would prepare contingency plans in case her worst fears proved justified.

As she gathered her thoughts, a realization struck her with unsettling clarity. She had been so focused on the Shard's properties that she'd overlooked something fundamental: her own fascination with it. The scientific detachment she prided herself on was slipping. Even now, part of her mind was already designing new experiments, imagining breakthrough papers, contemplating the neural mechanisms that might explain the impossible data, a quiet thrill that felt uncomfortably like the spark she'd seen in Devory's eyes.

Was this how it began for Devory? This subtle shift from professional curiosity to personal obsession?

She caught her reflection in the darkened monitor screen—the same focused expression she'd seen on Devory's face when he'd handed her the fragment. The same intensity in the eyes. The same certainty that they alone could unravel this mystery.

Mia forced her expression into professional neutrality as she left to meet Dr. Chen, the door sealing behind her with a soft hiss. The fragment was secure, the data encrypted, the evidence of her analysis deleted. No one would know what she had discovered.

At least that is what she told herself.

— 7 —
DOCTORS OFFICE

The door to Dr. Lilien Chen's office hushed shut behind Mia, its faint thud lost to the steady chug of the air recycling system. For a moment she stood there and gazed at Chen, who was busy studying something on her desk monitor. Chen didn't look up, she was totally absorbed.

Mia decided not to disturb her and instead cast her eyes around the room. As usual, she felt a pang of jealousy that Chen had gotten a larger space than she had been given, large enough to have an additional curtained-off cubicle for patient examinations, and a door that led to a private lab, where she performed pathology and post mortems. Unlike the sterility of her own lab, Chen's office had a more lived-in feel to it, a human touch, a testament to the hours spent there with the personal secrets of others.

The thing that always caught her eye was a shelf that housed a modest collection of medical textbooks, their pages worn from years of reference—prized relics in a world where digital records dominated. Among them, stood a framed photo of her extended family, a quiet tribute to her ancestry. It made Mia think of her own family, what little of it was left ...

Chen drew her eyes away from her monitor and straightened up, acknowledging Mia's presence. "We need to talk about the miners," she said, as Mia pulled up a chair and sat down.

Mia acknowledged Chen's concern, but guarded her response. "You're not wrong there," she said. "I've thought the same."

"I'm glad we're in agreement" Chen said, pulling up a biometric report on her monitor. "Ronald Hayes and Cornelius Dunbar—or Cory, as he likes to be called. Ice-vein miners, both assigned to oxygen extraction for the Landing Craft. I examined them yesterday after their shift. Their cortisol levels are sky-high, the kind of strain you'd find in soldiers during combat, not routine drilling. And their metabolism has spiked. Which is unusual. They're exhausted but wired."

Mia leaned forward, her scientific curiosity stirring despite her resolve to stay impartial. "Wired, how?"

"Elevated gamma amplitude, 40 to 50 Hz range. Hyperfocus, agitation. Hayes couldn't stop pacing; Dunbar's hands trembled like he'd overdosed on stimulants." A grimace briefly passed across Chen's normally unemotional features. "There's a rumor the chemistry lab's been synthesizing stimulants—amphetamines, maybe worse. Miners are pushing too hard. Someone's doping them up to keep quotas. Carson's Landing Craft deadline has got everyone stretched to breaking point."

Mia's thoughts snagged on the 40 Hz: again the same value. First the Shard, then Devory, and then her organoids. And now Hayes and Dunbar. "You know my opinion on the subject," Mia said maintaining her poise, "we're heading for burn-out, or worse, a serious accident. But something tells me that's *not* what's bothering you."

Chen knew Mia Hawthorn too well to not take a hint. "You want me to say it's the discovery of the Shard."

Mia raised her eyebrows suggestively.

Chen looked at her monitor briefly then returned her gaze to Mia. "I'll give you that. I think we can both agree the profiles don't fit. There's too much coherence across brain regions, not the intensification of the reward circuit pushing the rest of the brain. Something's off, yeah? You're the neuroscientist. You tell me."

Just like that. The professional gauntlet thrown down. Mia acknowledged her superior, knowing Chen had called her bluff. She thought of Devory's bright eyes, his sweaty hands, the Shard's impossible weight in her palm. And before she knew it, her resolve weakened, the words she wanted to hold back, they started tumbling out: "I've been studying something, something that might explain this," she began.

Chen's eyes slightly widened, but her face revealed that she was not completely surprised. "And?"

"What if it's not stimulants? What if it's something else …"

Chen's head tilted, a fractional shift. "Go on."

Mia faltered, the air stretched taut between them. "The Shard," she said finally, the word dropping like a stone—"I've managed to procure a fragment. I've been testing it."

"You did what?"

Mia held up her hand to stave off the inevitable recrimination. "I know. I know. Don't take this the wrong way. I was concerned about Devory."

Chen's expression darkened. "You have it with you?"

Mia hesitated. "Not here."

Chen's hands clenched ever so tightly over each other. "Then bring it to me. I need to study it."

Mia studied Chen's face, searching for any sign of hesitation. There was none. Just pure, clinical hunger. She exhaled. "Fine. But you have to promise me something first."

Chen narrowed her eyes. "What?"

"You don't tell anyone. Not Carson, not Kendricks, and especially not Prya Chandra." Mia leaned in. "This stays between us."

A pause. Then, slowly, Chen nodded. "Agreed."

Mia took a steady breath, then spoke with her hands. "The Shard's affecting neural activity, Lilien. Not just Devory but potentially anyone exposed to it. The miners, maybe even us."

"Is that so?"

Mia nodded. "I think it's rewriting their neural architecture at a fundamental level."

Chen crossed her arms, the overhead lights catching the silver streaks in her hair. "That's an extraordinary claim." Her voice was calm, but Mia saw the way her fingers nervously dug into her own flesh. "Show me the data."

Mia took her data stick from her pocket and plugged it into Chen's monitor. The files transmitted with a soft chime. "Look at the hippocampal formations first," she said as Chen's monitor displayed the images. "Then compare them to the prefrontal cortex activity."

The screens filled with rotating 3D models of neural scans, biochemical markers pulsing in rhythmic waves. Chen's breath hitched almost imperceptibly as she leaned forward, her reflection warping in the monitor's glow. A flicker of emotions passed over Chen's face—clinical curiosity, a hesitant pause, then a shadow of something more unsettling. Mia's skin prickled as she watched it unfold.

"This coherence ..." Chen murmured, enlarging a section of the scan. Her finger traced a cascade of synchronized gamma waves. "It's not just elevated activity. It's perfect phase coupling across all major networks. I've never seen neural oscillations like this before. They're so ... organized."

Mia shifted closer, her shoulder brushing Chen's. "Watch what happens at the 40 Hz frequency band." She zoomed in on the temporal lobe. "See how the waveforms don't just spike? They form standing wave patterns. Like the neurons are ..." She struggled for the right word.

"Resonating," Chen said, finding the word. "Like a quartz crystal under precise voltage." She turned sharply to Mia. "How many subjects show this?"

"So far just Devory ... and my organoids," Mia said. She pulled up another file. "The organoids ... their neural conduction length jumped from 6.5 mm to over 32 mm in under an hour. And Devory—" Her voice caught. "When I tested his reaction times yesterday, he was processing visual stimuli up to 100 milliseconds faster than humanly possible."

Chen's chair creaked as she sat back abruptly. The monitor's light painted her face in shifting violet blue shadows. For a long moment, she simply stared at the scans, her fingers steepled beneath her chin. When she finally spoke, her voice had dropped to a whisper. "This is incredible. But ... do you understand the mechanism?"

Mia expected the question and switched to a set of graphs. "I ran the Stern-Volmer analysis three times. The microtubule photoexcitation diffusion length exceeds all known parameters."

"Which means what?"

"They're harvesting more energy than they're capable of doing on their own."

A charged silence settled between them. The air recyclers hummed. Somewhere in the distance, a coolant pipe gurgled.

Chen suddenly stood and paced to the sanitizer station. She

pumped gel into her palms with unnecessary force. "We need more data. Full-spectrum blood work. Cerebrospinal fluid samples. I want biopsies from Hayes and Dunbar's temporal lobes—"

"Biopsies?" Mia's chair scraped against the floor as she stood. "Lilien, we're talking about invasive brain surgery on possibly compromised subjects. Prya won't stand for it."

"I'll convince her we're dealing with an unprecedented medical event," Chen said right away, her eyes glittering. "Do you realize what this could mean? If we can isolate the mechanism, we're looking at curing Alzheimer's, Parkinson's, maybe even reversing neural degradation entirely." Her voice took on a fervent edge. "Imagine a treatment that could extend healthy cognitive function by decades."

Mia's stomach twisted. She recognized that tone—it was the same one Devory had used in the Common Dome. Carefully controlling her emotions, she said: "At what cost? We don't understand the long-term effects. Hayes has already—"

"—Hayes was unstable to begin with," Chen interrupted. She returned to the monitor, her fingers dancing across the screen. "Look at these synaptic patterns. The efficiency gains alone are off the charts."

Mia yanked her data stick out of Chen's monitor. The suddenness of the action disrupted Chen's train of thought. Their gazes momentarily locked and Mia saw it—the subtle dilation of pupils, the faint tremor in her hands. The same signs she'd seen in Devory. It made her do something she had never done before;

assume superiority over Chen: "Listen Lilien," she said, "this isn't just research anymore. It's affecting you too."

Chen rocked back on her heels. "Don't be so dramatic. I'm simply recognizing an opportunity that—"

"—That what?" Mia pressed. "That justifies bypassing safety protocols? That makes it acceptable to experiment on cremates?" She gestured to the screen. "How do we even know it's not a contagion?"

The words fell like a guillotine between them. Chen's breathing had quickened, her chest rising and falling visibly beneath her medical tunic. For an instant, Mia thought she saw something flicker behind Chen's eyes; not just excitement, but something possessive.

Then Chen blinked, and her expression smoothed into clinical detachment. "We'll proceed methodically," she said, her voice carefully measured. "I'll handle the physical biopsies and blood work. You continue the neural modeling." She tapped her console, and the screens split into a dozen new data streams. "For now, all I want is for you to keep your promise. Lend me your fragment so I can get to work."

Mia opened her mouth to argue, but Chen's hardened exterior told her it was pointless. The unease in her chest tightened into a hard knot. The Shard wasn't just a puzzle to be solved anymore, it was a mirror, reflecting their deepest ambitions back at them in a twisted form.

Mia slowly got up and turned to leave. As she did so, Chen said,

"I'll be waiting here for you."

Mia nodded and headed to the door and pressed the touch screen to let herself out. For a fleeting moment, she caught her own reflection in the screen, and saw a pair of eyes staring back at herself. Who was that, she wondered? Was it the rational scientist, or something else the Shard had already begun to create?

She shuddered and quickly walked out, the door closing behind her.

— 8 —
THE ARCHIVES

After reluctantly handing her fragment over to Chen, Mia headed for the Archives. She passed through the Common Dome and headed to the "east" sector. The Archives had one of the largest entrance doors on the GenStar, braced by a reinforced metal arch. She pressed the intercom, identified herself and waited for the door to open. The cavernous space seemed to lean inward, its vaulted ceiling strung with flickering full-spectrum lights that cast darting shadows across data towers and relic cases. Her nose tickled from the musty odor in the air. Somewhere in the gloom, a ventilation duct sputtered, like the rattling of an old man's bones. In the background, Charlie, Tiber's humanoid simulant worked diligently among the stacks, his 1.75 meter tall alloy frame moving with quiet precision as he updated the cataloging system, his LED eyes glowing a faint amber in the dimness.

Tiber Solis sat hunched at his workstation, his frail silhouette illuminated by the blue glow of a dozen screens. He didn't turn as Mia approached, but his voice—surprisingly deep and mellifluous for his gaunt appearance—demanded attention: "I see you've brought me a caseload of interviews to file ..."

Mia approached his workstation. "Not interviews. Evidence."

"Evidence?" His chair squeaked as he swiveled to face her. His sunken cheeks accentuated his eyes, which seemed to have grown in size and brightness since she last saw him. "You look worse than me," he said. "What have you been doing?"

"I've been trying to find answers."

He glanced at Charlie, who stepped forward silently from the shadows, an antique book in his hands. Its leather binding cracked, its pages yellowed. Charlie handed it to Tiber with a slight nod, then receded back to his work among the stacks, his presence fading into the background hum. Tiber set the book down with a faint thud. "Answers ..."

There was an empty chair nearby, its cushioning worn down to the metal. Mia grabbed it and sat down without fussing. She met his gaze, undeterred. "Is that what you're doing? Looking for answers?"

Tiber indicated the book. "The Collapse of Complex Societies by Tainter."

"I see. The great human flaw: we never learn from our mistakes."

"So you *have* been listening to me."

Mia let a faint smile cross her features, revealing a part of herself she usually only reserved for those she genuinely trusted. It was no secret Tiber had a soft spot for her. If he only knew how much she loved him—not like she loved Devory, but deeper, like a catfish lurking beneath the mud, always ready to feed off whatever morsel he threw her way. "At least there's someone on this damned rock that appreciates my walnut-sized brain."

"It's not the size that counts ..." He opened the book and quoted: "The straw that breaks the camel's back is usually absurdly small. A grain shortage here, a corrupted official there. Systems under tension require perfect balance. Drop a pebble into the pond ..."

"And the ripples spread out."

"Precisely." Tiber snapped the book shut. "Tell me why you're really here. And don't insult me by pretending it's a social call."

Mia exhaled sharply. She pulled the data chip from her pocket, the one containing the Shard research, and handed it to him. "I found our pebble."

Tiber flicked a concerned look at her, then jacked it into his personal monitor. Mia knew she could trust him. Over the years, he had forged an arsenal of custom tools, cryptic locks and digital snares, to shield his archives from prying eyes and data thieves. No one breached the Grand Archivist's fortress.

"Ok," Tiber said, scrolling through the data, "what have we got here?"

Mia walked him through it—the neural patterns, the organoid changes, Devory's accelerated cognition. As she spoke, she watched Tiber's eyes darken, his fingers unconsciously tugging at the lanyard that held his ancient data stick, a reminder of the unbroken chain.

When she finished, he said, "So it's worse than I thought."

She was afraid he would say that.

Tiber fell into thought for a while, clearly wrestling with something. He dropped all pretense and said, "There's something I need to tell you."

"What?"

"Prya came in this morning. Peppering me with questions about the Shard."

Mia felt her whole body tighten up. "What did you tell her?"

"The truth, of course."

"Which is?"

"There hasn't been anything like it in all our history."

"How did she take it?"

"It excited her."

Mia's stomach dropped. "No ..."

"Yep." He let that fact sink in for a moment. Then added, "Of course, I didn't need to remind her of the Long Night."

"The Long Night ..." Mia murmured.

Tiber nodded. "Few talk about it. It started with the Spiritual Advisor of the time—Tenzen Marduke. Revered. Powerful. Until he was exposed as a child predator." He said it as a matter of historical fact, but his fingers curled tightly against the table. "The revelation shattered everything. The faith that bound the crew together disintegrated overnight."

"I can imagine. And the fallout?"

"Mass suicides. Riots. Families who had built their lives around the Great Continuance lost their anchor. Some renounced it completely. Others doubled down, blaming the victims, claiming it was some sort of test. The division nearly turned into outright civil war."

"How did it end?"

"The way these things always end." Tiber's eyes flashed. "With blood. The ones who tried to move forward without faith were driven into exile. The remaining faithful elected a new leader, one who promised to cleanse the past and restore order. If you ask me, it was classic case of cognitive dissonance: when reality doesn't fit the story people tell themselves, they either change the story or destroy the evidence."

Mia couldn't help smirking. It was a term she had taught him.

Tiber said, "As for Prya, she's learned her lesson well. She understands better than anyone that control over belief is control over people." He leaned back, tapping the desk. "That's why she holds the Ark Festival. She can anoint the objects people donate as sacred and no one will question her. She's reinforcing the Great Continuance, because she knows what happens if it crumbles."

Mia nodded in understanding. It was a familiar refrain. So long as the illusion of control was maintained, everything continues as normal. What ever that was. Her gaze drifted to the book on Tiber's desk. Clearly, the past wasn't just a lesson. It was a warning. "And if she gets her hands on the Shard?" she dared ask.

Tiber exhaled slowly. "She'll try to sanctify it. Make it proof of divine intervention. My worry is what happens if she doesn't get her way. She's conniving and liable to split everyone into factions. She could just as easily turn on it, and then who knows where this thing will go."

Hearing Tiber say it, he wasn't one to exaggerate. She believed him. She realized she had been thinking too small. The whole ship was now involved.

And then the thought hit her: what if she was already affected by it? What if her judgment was already clouded? She shuddered to think of that possibility. Pushed it out of her mind, quarantined it like an infected animal.

Tiber was looking at her. He didn't say anything. But she knew he understood.

It seemed like they were standing on a precipice. The slightest mistake now, and all their lives could be put in jeopardy. Or, the whole thing could turn out to be nothing more than a figment of their imaginations. How she wished that would be true.

But before either of them could break the spell, the ship's comms crackled to life. Prya's voice, honeyed yet congealed with something oddly unsettling, echoed through the Archives:

The Eighth Reflection foretold a sign to guide us. Do you recall its promise? A marker for the journey's end, a light in our long darkness. And I tell you, it begins right here, right now. We shall rise from the shadows we've hid amongst for too long. Have you not faltered in your watch? Have you not strayed from the path laid before us? Repent your darkness! Join me in the Observatory to claim the dawn of our redemption!

Tiber looked past Mia, through the walls at the belly of the ship. "I guess we're in for it now."

Mia felt the blood drain from her head. She lurched to her feet, the room tilting around her. She had to move. Now.

She headed for the exit.

Tiber yanked her data stick out of his monitor and waved it at her. "You forgot this!"

But Mia wasn't listening. She was already running.

— 9 —

RIOT

The Common Dome buzzed like a live wire, carrying a fervor that rivaled the Ark Festival—maybe even surpassing it—yet, unlike the festival, there wasn't a drop of alcohol in sight. Arkonauts packed in tight, shoulder to shoulder, drawn by whispers of the Shard like moths to a flame. Below, Eridani's radiance seeped through the inverted glass Dome, bathing their upturned faces in a holy shimmer. Prya Chandra commanded the center, standing atop a stout dining table, her purple robes flowing around her agile body as she slowly rotated to show everyone what was in her hand: a glass vial with a fragment of the Shard.

"Behold!" she cried, thrusting the vial skyward. The fragment caught Eridani's light and seemed to gleam more brightly than usual, its black planes drinking the light yet radiating a promise that prickled the skin. For an instant, Mia swore the glow pulsed—like a heartbeat. A trick of refraction. Or something else.

"The divine has spoken!" Prya's voice rang out. "This is no mere rock! It is the hand of destiny, reaching across sixteen thousand years to anoint us!"

The crowd leaned in, eyes wide, breaths held. Engineers abandoned their central tables, agricultural workers drifted from the hydroponics edge, and even the navigation crew turned from their starward vigil. The vial became a lodestone, drawing them into Prya's orbit. Even from the entrance, Mia could see faces growing taut, hands clenching and unclenching, a current of agitation just waiting to break out.

Mia slipped through the entrance, her breathing quickening as she took in the scene. Having come straight from the Archives, Tiber's warning still ringing in her ears, this … this was a storm she had not foreseen. The Common Dome, once a refuge of routine, now seethed with a fervor more intense than any Ark Festival. Prya's sermon rolled over the packed listeners, her voice a siren call drawing them into a volatile trance. Mia edged along the eastern curve, shadows her only cover, until her gaze landed on Rivel Torvax near the front.

Rivel stood with arms crossed, her single eye fixed on Prya with a philosopher's fervor—less devotion than obsession, as if the Shard were an equation she needed to solve. Her head tilted, her smile faltering—just for an instant. A flicker of unease crossed her face, quickly buried beneath a mask of calculation. But Mia saw it.

Prya still held the vial aloft, its fragment glinting, and a jolt hit Mia: Rivel must have given it to her. Who else could pry a piece from the Shard, with Devory buried in Mining Control? Rivel's hammer, her 40 Hz trick, no doubt the culprit. Mia's stomach churned. She'd handed Devory's fragment to Chen, and now Rivel had fed Prya's fire.

Prya's voice soared, her free hand slashing the air. "The Continuance promised us a new dawn! Why grovel for a planet when we can become more? This is our rapture!" She shook the vial, and gasps rippled through the crowd, their eyes reflecting the fragment's impossible refraction.

Rivel's brow creased deeper. Mia watched tension gather in her shoulders, a sign perhaps that something was off. Prya pressed on, her tone a thunderclap: "Eridani-B was never the goal! The Shard is our salvation!"

The crowd roared: "YES!" Their hands clawed upward, faces contorting with something between ecstasy and violence.

Rivel's jaw tightened, her arms dropping as her eye narrowed to a slit. Mia saw it: Prya's words had crossed a line. This was not what Rivel believed. She clenched her fists, then shoved forward, parting the throng, her braid snapping like a whip as she stormed toward the exit.

Mia made an instant decision. She darted after Rivel, weaving through the crowd as Prya's voice boomed in her wake: "The ancestors wept so we could rejoice! The Shard is their will!" Chants of "SHARD! SHARD!" chased her into the corridor.

Mia reached Rivel and grabbed her shoulder. Rivel spun, fist half-raised—then recognized her. A heartbeat passed between them. "She lied," Rivel hissed, her voice raw. "I gave her that piece thinking she saw potential—not some damned idol."

Mia caught her arm. "Potential for what?"

Rivel's eye flashed. "To evolve, not grovel at the feet of some rock! Extend our consciousness beyond these puny confines." She yanked free, trembling. "I was a fool."

Mia nodded. Rivel's hammer, her faith in science over scripture—"We need to—"

A roar from the Dome cut her off. Mia glanced back. Prya's sermon had hit a fever pitch, and chaos was breaking loose. Her instinct screamed: Find Carson. Only he could stop this before it exploded. She scanned the crowd—Seong-Min Park stood rigid near the back, Phil Polsen whispering beside her—she spun on her heels and headed back to warn them, pull them out, not sure if it was fear or courage that drove her.

And then, as if her thoughts had been read, Captain Frederick Carson stormed through the entrance, his blue uniform a blade slicing through the madness. Martin Kendricks trailed behind him, Lora Vasquez at his flank, her black uniform with red accents mirroring his as they moved in lockstep. Carson's eyes locked on Prya, still atop the table. For a heartbeat, he stood frozen, unable to reconcile the scene—the people he swore to protect, faces twisted in fanaticism.

Kendricks acted first. He shoved past Carson, shouting: "Chandra, get down!" Vasquez surged alongside him, her movements fluid and precise, a silent shadow amplifying his authority.

Prya ignored them, serene. "The Shard has chosen us!"

Kendricks vaulted onto the table, boots slamming down hard. "Enough!" He snatched the vial from her grip—the crowd gasped—then surged. Vasquez leapt into the fray below, her baton flashing as she intercepted a lunging figure, driving them back with a practiced strike. Prya's adherents, wild-eyed, lunged for the relic. "It is ours!" one of them shrieked, a woman from the canteen.

The Dome erupted.

Fists flew. Tables overturned. A hydroponic crate tipped, spilling nutrient-rich water across the floor as bodies grappled. A worker screamed as she was trampled. Carson took a punch to the ribs but held his ground, striking back with military precision. Kendricks tackled a man lunging for the vial in his hand—they crashed into a table, dishes exploding in a hail of ceramic shards—while Vasquez hauled another assailant off him, her grip iron-strong as she pinned them to the wet floor. Security officers poured in, batons cracking against flesh, the sickening thud of impact drowned by the frenzy. A bloodied hand scrabbled for the Shard, fingers slipping on wet flooring before Vasquez's boot crushed them down.

Another man, a miner, succeeded in knocking Kendricks

off balance. The vial slipped from Kendricks' grip, shattering. A pneumatics technician dove for the fragment, raising it triumphantly—until Carson's fist shattered his jaw. The man crumpled. Carson seized the shard, his knuckles split and bleeding.

Mia met his gaze across the wreckage. He palmed the fragment, its edges biting into his bloody fist, and jammed it into his pocket. "This ends here," he muttered, wiping his hand on his uniform, leaving a red smear across the blue. "Get Chen. We need answers."

Mia nodded, though her mind spun. In days, everything had unraveled. How? Below, in the bowl of the dome, 82 Eridani glared through the glass, indifferent. Three months to orbit, and the GenStar was fracturing—not from gravity's pull, but from the weight of human minds snapping under the Shard's influence.

– 10 –
REASSESSMENT

The Command Center's air hung thick with the metallic tang of blood and ozone, a residue from the riot that had seeped into the vents and clung like a second skin. It had taken Kendricks and Vasquez several hours, with help from security pulled from other sectors, to secure the Dome and rig a makeshift brig for the worst offenders. Most were miners, a fact that galled Carson. It was another blow to oxygen production. He stood at the head, still unwashed from the fight, his jumpsuit stained with grime and blood, marking the cost of the day. His steely gaze locked onto Prya across the table as he pulled the fragment from his pocket with a lacerated, bruised hand, holding it up briefly for all to see before setting it down. He stared at it as it lay perfectly still on the surface, unable to understand why this was odd, forgetting that the table was non-metallic. But the thought quickly passed as he, like the others, found himself entranced by its black trapezoid shape, its edges too perfect, its light-swallowing surface blacker than the cold recesses of space. With great effort, he pulled his eyes away and fixed them back on Prya.

"Explain," he said to Chandra, his voice low but sharp with authority. "No sermons. No poetry. Just tell me—what exactly did you think would happen when you held that thing up in front of your followers?"

Prya Chandra sat perfectly still. Her violet robes remained pristine despite the wreckage around her. The geometric patterns along her sleeves shimmered faintly in the flickering light—woven with something finer than thread. She didn't look away from Carson.

"The Continuance texts are clear, Captain," she said calmly, her tone like polished marble. "They speak of a sign that appears at journey's end. Not from Earth, not from heaven, but from the in-between. When the miners uncovered the Shard, I knew. I *just* knew. It doesn't obey natural laws. It resonates. It's calling something in us."

Carson's fingers twitched toward the fragment but stopped short. "What do you mean by 'resonates'?"

"You've seen the data," Prya said, voice smooth. "It repels metal tools. It absorbs radiation without heating. It has a density that shouldn't be possible. But numbers only tell part of the story.

When I held it during the sermon, I felt it ... like I was waking up from a dream I didn't know I'd been having."

Carson's voice turned harder, more direct. "And the riot? Was that part of your awakening too? The concussions, the broken bones, the lawlessness?"

Prya exhaled slowly, the sound barely audible. She eyed Chen warily, noting the blood stains on her uniform, a sign of the numerous cuts and lacerations she had to treat. A job she couldn't have accomplished without the help of her nurse, Jacqueline Melange, who was still tending to the injured as they spoke. "Growth isn't always gentle, Captain. The Shard doesn't hide what we are. It shows us. And some people—" she paused, "—some people fight that truth. They'll cling to old illusions until they break."

"That's not an explanation," Carson said, leaning over the table. His shadow fell across the fragment like a shroud. "You threw a match into a powder keg. I need to know why."

She shifted slightly, the first crack in her composure. "Because they deserved to know the truth," she said, her voice gaining edge. "We've spent sixteen thousand years telling ourselves the void is empty. It's not. The Shard proves it. And if that truth breaks a few ribs on the way in, so be it."

A beat passed. The fragment held the center of the room, taunting them with its mystery.

"Bullshit."

Kendricks' voice cut through, rough and tired. The security chief stood stiff in his black uniform, red accents smeared with grime. His electro-baton—a clear violation of Regulation 1147—hung visibly at his hip, a not-so-subtle threat. His eyes were bloodshot, his face drawn from hours of roughing unruly prisoners.

"It's a rock, Chandra," he said, stepping forward. "Yeah, it's weird. I'll give you that. But it's just a rock. And it's messing with our heads. Captain—I already made my recommendation, if anyone cares to remember. I said it then, and I'll say it now: dump it from the airlock. No ifs and buts."

Rivel laughed, high and off-kilter. She whipped around to face Kendricks, her single eye catching the light like a lens flare. "Of course that's your solution. Anything you can't lock down, just toss it out." She leaned forward. "Don't you see? It's showing us what our true potential could be. How it does it? I'm not sure yet, I'm still doing calculations, but my own thoughts ..." She tapped her temple. "Clearer, sharper, like I've been squinting my whole life and now I've just opened my eyes. That's not decay. That's evolution in real time. And we can either lean into it or spend the next century rotting here, wondering what might've been."

Kendricks' hand drifted toward the rod. "Open your eyes, Torvax. Three people in medbay with neural bleeding. Six more locked in the brig because they tried to claw their own skins off. However pretty your theories are, that's the reality we're living right now."

"It's potential," Rivel shot back. "Evolution doesn't come clean. It's not neat. You want a guidebook? There isn't one."

Prya narrowed her eyes. "Faith built the Continuance, Torvax."

"Faith?" Rivel's voice dropped, almost a growl. "Faith's what people use when they're afraid to look deeper. The Shard's not divine. It's something else—something we can understand. That's the point."

Dr. Chen's voice slipped in, cold and precise. "She's not entirely wrong."

Everyone turned.

"I've been analyzing the fragment Mia had," she said, voice clinical but tinged with something electric. "Devory's reaction time improved by over 150 milliseconds—beyond known human capability. Microtubule conduction in the organoids is five times normal. This isn't theoretical, it's observable. And more than that? It's regenerative. Do you realize what this means? We've spent generations watching minds unravel. Alzheimer's, Parkinson's, the slow rot of time. And now? The organoids show regeneration. Not just stabilization, but a reversal. If we can harness this, really understand it … mortality itself becomes negotiable."

Mia glared at Chen ferociously. "That's not what the data says," she snapped. "You're skipping steps, Lilien. We don't even understand the mechanism. And we sure as hell don't know the cost."

Chen's smile was thin and tight. "You had it for days. You didn't share. Why?"

A wave of murmurs swept the room. Kendricks muttered a curse, then added: "You think I wouldn't notice?"

At the same time Carson said, "Is it true?" His voice was quiet but hard as steel.

Mia reacted with insolence, her fists tight at her sides. "I didn't have it for days. That's not true. I got it yesterday ... I kept it for analysis," she said. "Everything happened so fast, I didn't have time to think. Because something is happening. To all of us. Devory, Hayes, the organoids, it's the same signature. I didn't want to rush in blind. I needed to know what we were dealing with."

"You disobeyed a direct order," Carson said, each word deliberate. "After the Council meeting, I made it clear: No one goes near the Shard. Our top priority was oxygen extraction."

"I know," Mia said, voice rising. "But this thing ... it's changing us. The coherence at 40 Hz isn't random. The microtubules are processing data at a faster rate. I'm still trying to find out why."

Carson's eyes narrowed. "Exactly how is it changing us?"

Mia took a breath. "That's the question I've been trying to answer. I don't know."

Chen interjected, voice ringing clear. "What Mia is saying is the Shard is enhancing us. That can only be a good thing. I bet we can turn it therapeutic use. I suggest we put it through clinical trials. This could be the answer we've all been looking for."

Mia turned on her, furious. "Clinical trials? Excuse me, but we don't even know if it is safe yet. To suggest that we should start clinical trials without knowing its effects is reckless!"

"It's not reckless," Chen snapped. "It's real. Devory's latency dropped to a level that shouldn't be possible. That's not a fluke. That's actionable evidence."

Polsen was nodding in agreement. "It might be a bioelectric field," he offered enthusiastically. "You just said the Shard is altering the electrical conductivity within microtubules. If true, then this effect might depend on the strength of the field."

Mia's eyes widened, her mind racing. Bioelectric. Quantum coherence. It fit—Devory, the organoids, the coherence at 40 Hz.

"It's plausible," she said. "If the microtubules are processing more than they should ... it could explain the cognitive shifts. But we still don't know the range, or the long-term effects."

"Then we test," Chen said sharply. "No more delays. I can solve this."

Anna Teoh spoke up, quiet but firm. But it was completely unexpected. "Mia had to study it."

Aldo Vandergamma turned toward her. "What the hell are you saying?"

"If it's altering cognition," Anna said, tracing an arc in the air, "then it could be skewing our calculations. Orbital burns. Navigation trajectories. We can't just ignore that."

Aldo scoffed. "It's not affecting navigation."

"Are you sure?"

Tiber coughed to get everyone's attention. He waited till he had it. "I promised myself I wouldn't lecture," he said. "But we're past that now. How many of you here know why Rome fell?"

An awkward silence washed over the room.

"What's Rome?" Anna asked meekly.

Tiber turned slowly toward her. His eyes, pale and clouded,

seemed to stare across centuries.

"Rome was a civilization," he said, voice even, "the most powerful the Earth had ever known. Its armies spanned continents. Its engineers built aqueducts that carried water for miles. Its laws shaped the very idea of justice. For nearly a thousand years, it believed itself eternal."

He looked at Prya as he said 'eternal,' the hologram's amber light pooling in his weathered features.

"But it wasn't gods or barbarians that killed Rome," he said, turning back to Anna, "it was something slower. Something internal. Corruption. Hubris. A refusal to adapt. They forgot what had made them great, their discipline, their unity, their skepticism, and replaced it with spectacle, with zealotry, with blind faith in their own invincibility."

The room was still. Even Rivel, twitching a moment ago, now stood motionless.

"They stopped questioning. They let the voices promising rebirth drown out those warning of decline. Cults rose. Rational thought decayed. And when the center no longer held, when the legions stopped marching and the aqueducts ran dry ... it didn't fall in a blaze. It crumbled. From within."

Tiber looked directly at Prya, then at Chen.

"We are not Rome," he said. "But we are just as mortal. Just as proud. Just as susceptible to visions we don't understand. That Shard—" he pointed at the black trapezoid—"it may be our aqueduct. Or it may be our Coliseum, flooded with blood for the sake of belief."

He turned back to the group.

"So ask yourself: Are we building something? Or worshiping it? Because one path leads to a future. The other leads to ruin."

Carson rubbed his temples, exhaling slowly. "We quarantine it. Full isolation. Everyone exposed gets tested. Mia, Chen—you lead the process. And from this moment forward, no more sermons. No more unsanctioned research."

Prya opened her mouth—

"That's an order," Carson said, final. Then more quietly, he reached his hand over the fragment. "This isn't just a rock," he said quietly. "It's an anomaly, like Tycho's star, its forcing us to update our model of the cosmos. We can accept it or deny it at our peril."

Murmurs spread. The Captain's analogy with Tycho's star, the event that broke 2000 years of Ptolemaic astronomy was apt. But just how much did they understand? How much were they capable of understanding?

Mia watched as Carson's bruised and battered hand hovered above the fragment, its black edges catching the diffused light.

If any of their theories held water, this wasn't just a test of science, it was a reckoning of will.

And if there was one thing she knew about human will, it could be easily warped. Or it could just as easily be stubborn to a fault.

But the way Carson stared at it, she got the distinct feeling that this was more, that the Shard was reconfiguring him into something even more terrifying.

She hurried out of the room, blocking the vision from her mind. Stop it! she told herself. Stop it! You are a neuroscientist. It's all in your brain!

But the more she said it, the less she believed it.

– 11 –
FAILSAFE

Mia lay sprawled across the narrow bunk in her quarters aboard the GenStar, the thin mattress sagging under her weight. Three months remained until orbital insertion around 82 Eridani, yet the ship's chaos churned in her mind like a relentless engine. Her quarters were a sealed pocket deep in the asteroid's superstructure, its rough walls coated in smooth polygrout, a polymer-regolith blend that gleamed faintly under the dim overhead light, insulating her from the raw stone beyond. Stripped down to her underwear, she pulled the coarse blanket up to her chin, staring blankly at the ceiling.

Sleep beckoned, an obliteration she craved to drown out the storm of her thoughts, but her mind refused to quiet.

"Forty hertz … microtubules … coherence spikes," she muttered, the words looping like a broken recording. The Command Center's reckoning gnawed at her—Chen's wild-eyed talk of immortality, Prya's unshaken faith, Kendricks' bloodshot rage, Carson's bruised fist closing around the Shard. And her own betrayal, laid bare: hiding the fragment, testing it, lying to them all.

"Am I even me anymore?" she whispered, her voice fading as exhaustion tugged her toward the edge of oblivion.

Her eyelids fluttered, heavy and slow, as the hum of the GenStar's vast systems blended into a dull lullaby through the sealed walls. She teetered on the brink of sleep, her body sinking into the bunk, when a sharp chime jolted her awake. Her heart lurched. She bolted upright, the blanket slipping to her waist.

The glass touch panel embedded in the wall beside her bed flickered with a feed, showing Charlie's generations-old form outside her door—battered titanium panels dulled and scratched from centuries of use. His synthetic voice crackled through the speaker, clipped but familiar, his worn silicone face shifting as the lower lip moved.

"Mia, it's me. Can I come in?"

She groaned and flopped back onto the pillow, her hair a tangled mess across her face. "No, Charlie. I'm trying to sleep. Go away." Her voice was hoarse, edged with irritation and fatigue.

The panel chimed again. His tone was insistent but gentle. "I'm not leaving, Mia. Let me in—just for a minute. Please."

She glared at the screen, tempted to ignore him, but the spark of his persistence pierced her haze. With a sigh, she reached over and tapped the release icon. The door slid open with a hiss, admitting the simulant robot. His body, a patchwork of dull titanium panels dented and faded, clicked softly at the joints as he stepped inside. His camera eyes, glowing a warm amber with infrared and ultraviolet range, scanned her disheveled form, while his silicone face creased faintly into a tilt of concern.

"You look like you've seen the end of the universe, Mia," he said, lower lip twitching with each word. "Spoiler: it's still 92 days away."

Mia pulled the blanket back up to her shoulders, her glare softening into exhaustion. "I was almost asleep, Charlie. What do you want?" She rubbed her eyes with the heels of her hands, fighting the urge to curl back into her cocoon.

Charlie settled onto a stool beside the bunk with a faint creak, his movements precise yet stiff from age, a relic mimicking life. "You're Mia Hawthorn, neuroscientist, resident skeptic, and—judging by that blanket cocoon—someone who's really committed to hiding." His camera eyes flickered playfully to blue, though his tone softened, silicone face creasing faintly. "I'm not here to drag you back to the chaos. Just talk to me. What's got you buried in here like a hibernating animal?"

She laughed, a brittle sound that cracked in her chest, and pushed the blanket down to her waist, sitting up slightly. "Did Tiber send you?"

"He's worried about you."

"I'm worried about me."

"Talk to me, Mia."

"It's the Shard, Charlie. It's affecting all of us."

"Tiber mentioned it, but I don't understand. It's inorganic. It's a dead piece of matter."

"Yes," Mia agreed, "but it's doing something to us."

"Okay, let's assume it is."

She slumped back against the polygrout wall, her voice dropping to a raw whisper. "It's my own fault. I allowed myself to be exposed to it."

"But you needed to test it. You were doing your duty."

Her hands pressed against her temples, fingers digging in as if to anchor her fracturing mind. "I keep seeing patterns. Forty hertz echoes, coherence spikes, microtubules acting like they've got a mind of their own. I'm a scientist, Charlie. I'm supposed to trust my data, my reason. But what if it's all in my head? What if I'm not me anymore?"

Charlie's camera eyes shifted to a steady amber, his battered frame still as he processed her words. Then he leaned forward, his voice calm but firm, lower lip shifting. "You're still here, Mia. Maybe a little rattled, but here. If you're scared of losing yourself, let's make sure you don't. Let's create a failsafe. A memory anchor. Something to pull you back if things get blurry."

Mia blinked. The idea cut through her fog like a faint light in the dark. "A recording?" she asked, propping herself on an elbow as the blanket slipped further.

Charlie nodded, his eyes brightening to blue. "Something immutable. A tether to who you are, not what the Shard might make you think you are. Do you agree?"

She exhaled, long and shaky, and nodded, sitting up fully, the blanket pooling around her hips. "Okay. Let's do it." Her fingers brushed the edge of the bunk, grounding her against the cold polygrout floor. "I just… I need to know I can come back."

Charlie tilted his head, eyes shifting to a curious blue. "Alright. Recording initiated. What's the anchor? What's the core of Mia Hawthorn we're locking down?"

Mia hesitated. Her gaze drifted to the smooth polygrout wall, its sealed surface a quiet barrier against the world beyond. "My value system," she said finally, her voice firmer despite the lingering exhaustion. "It's what drives me. What keeps me … me. If I lose that, I'm gone."

Charlie's eyes flickered, processing. "Okay then. What's your highest value?"

"Consciousness," Mia said without hesitation, her eyes locking onto his. The word landed like a vow.

Charlie's stared at Mia, thinking. Then said, "Consciousness. Why that?"

Mia leaned back against the polygrout wall, her anxiety easing as conviction took hold, though her body still ached for sleep. "Think about it, Charlie. In all of human history—sixteen thousand years on this ship, millennia before that—we've found no trace of another intelligent civilization. No signals, no ruins, nothing. We might be it. The only ones who can look at the universe and see it looking back. If human consciousness dies, the universe loses its mirror. Its ability to know itself. That would be the greatest loss imaginable, and therefore it has the highest value."

Charlie's eyes pulsed softly. "So, you're saying we're the universe's self-awareness?"

"Exactly," Mia said, a faint smile tugging at her lips. "Without us, it's just stars and silence."

Charlie tilted his head again, his silicone face creasing into thoughtfulness. "Alright, I'll buy it. What's your second highest value?"

"Fantasy," Mia said, inhaling deeply then exhaling again. She felt herself starting to loosen up.

"Fantasy?" Charlie's tone rose, eyes flashing with interest. "You're gonna have to unpack that one."

Mia smile for the first time since the meeting. This was the first conversation she had had in a long time that made her feel grounded. "It's the raw stuff of consciousness Charlie," she said. "Wild, unshaped. Every leap forward, every invention, every story we've told since we had language started as a fantasy. It's the source of creativity, knowledge, understanding. Without it, we'd still be banging rocks together."

Charlie reflected on that, not completely convinced. "But how does fantasy become knowledge? It doesn't seem to logically follow."

Mia's eyes glinted, a spark of her old self breaking through. "Ask me my third value."

Charlie's eyes pulsed a deeper red this time. "Alright, you've got me. What's your third highest value?"

"Science," Mia said confidently. "It's the great filter. Consciousness spits out fantasies. Science takes them, tests them, then spits out facts on one side, and discards the rest as magical thinking."

Charlie's eyes brightened. "I can't argue with that," he said. "It's the one big difference between my brain and yours. I'm not judging. I'll never come up with something quirky—truly surprising. But somehow, I don't think humans would be too happy if I was programmed to do that."

"No," Mia said. "We want you to be logical and boring."

Charlie's face crinkled into a half grin, his best. "So what's the fourth value. I'm really curious now."

"Evolution."

Charlie nodded. "Consciousness, fantasy, science, then evolution. I suppose it makes sense."

"Nothing makes sense without evolution Charlie."

"True. We all evolved. Although I was built."

"You *have* evolved Charlie. You're hundreds of years old. How many versions of you do you think there are?"

"Well, version 1.0 when I was just—"

"Charlie—"

"Yes?"

"I don't need to hear every version of yourself. I like you just as you are now."

Charlie's camera eyes shifted to a warm purple, a rare hue she'd only seen when he was deeply engaged, his silicone face rippling faintly. "Alright, Mia. I've got it. Consciousness—the universe's mirror. Fantasy—the spark of creation. Science—the filter for truth. Evolution—the web that weaves life." He paused, voice softening. "That's your anchor. Recording terminated."

Mia's gaze locked onto his, the weight of this commitment balanced against the fragility of the moment. "That's me, Charlie. That's my core." She reached out and gripped his dull titanium arm over the blanket. "Promise me—if I lose myself, if this thing takes me too far, you'll show me this recording. You'll bring me back."

Charlie's amber eyes held steady, his hand unwavering. "I promise, Mia. I'll keep it safe. It's locked in my memory banks." He let go and tapped his chest to emphasize their extraordinary pact.

– 12 –

EMERGENCY

Mia didn't hear the door open. She had passed into sleep some-time in the pre-dawn hour, wrapped in her blanket, her limbs tangled in the cooling folds of fabric, body slack and mind silenced for the first time in what felt like weeks.

She only stirred when she felt the dip of the mattress.

A shape slid in behind her—hesitant at first, careful not to wake her. But his presence was unmistakable. The arm, lean and warm, slipped around her waist. The familiar scent—part grease, part ozone, something mineral like glacier dust—settled into her hair.

Devory.

For one vulnerable moment, she let herself believe everything was normal. That nothing had gone wrong. That there was no Shard. No riot. No neural contamination. Just this quiet warmth, this shared breath in a sealed room where no one could reach them. He didn't speak. Didn't move beyond the instinctive curl of his body against hers, exhaustion emanating from him like radiation. She could feel it in his muscles, the way his breathing dropped into that thick, dreamless cadence of the truly spent.

He was exhausted, of course, he'd been running the electrolysis rigs nonstop, managing oxygen reclamation from the ice veins near Shaft 16. And still, he hadn't stopped searching. Every off-hour spent combing tool manifests and mineral carts, chasing down the faintest hints of additional Shard fragments. Mia knew. She'd seen the guilt in his eyes, the obsession lighting up whenever she managed to catch him unawares.

His increasing use of sleeping pills worried her, but at least it knocked him out, let him sleep. There would be a time, in the near future, when she would have to confront him about it. But not now. She would grant him this moment of oblivion. Not just for him, but for herself. She wanted to soak in this echo of the man he used to be. To believe, just for a few more minutes, that he was still Devory, and she was still Mia—a scientist, a partner, not a warden of her own lover's crumbling mind.

She let that thought wash over her, and she began to sink back into the welcoming darkness.

—Until the chime shattered it.

Ping.

Ping.

Mia stirred, groggy. Then the voice came through the wall panel, clipped and urgent.

"Mia, it's Chen. We have an emergency. Are you awake?"

Her eyes opened fully. "Yeah," she said, voice hoarse. "What's going on?"

"Ronny Hayes has gone off the handle. Lora says he killed a sex worker, and now ... he's taken another as a hostage. He has a knife."

"Shit." Mia threw the blanket off and leapt out of bed. She grabbed her coveralls off the chair and hopped into them like a fire had started under her.

"Kendricks has threatened to go in there guns blazing," Chen continued. "We need to talk Hayes down. Otherwise we'll have to deliver not just one, but potentially two or more bodies to the recycler."

"Copy," Mia said, zipping halfway before turning back toward the bed.

Devory had stirred, blinking up at her, hair mussed, eyes slow to focus.

"What's going on?" he asked, still half-lost in the dreamspace between wakefulness and fugue.

She finished zipping up.

"Wha—what is it?"

She tapped the control panel; the door hissed open.

"Mia!"

"It's the Shard Devory. It's the Shard."

And she was gone.

The door sealed behind her. The room returned to silence—except for the fading warmth on the mattress and the waft of Mia's scent still in the air. Devory stared believing it was a dream, his eyes searching the space where she'd just stood.

– 13 –

PLEASURE QUARTERS

The Pleasure Quarters smelled of sweat and cheap perfume, a dim corridor lit by flickering garish red panels that buzzed faintly, like dying insects. Lora Vasquez stood rigid outside a sealed door, her black-and-red uniform stark against the scuffed polygrout wall. Her earpiece crackled with static, her voice steady but tight, projecting through the door's comms panel.

"Stay calm, Hayes. We can work this out. Just talk to me."

Inside, Ronny Hayes was unraveling. Lora had heard the screams thirty minutes ago, sprinting from her sentry post before the second girl's alarm even registered. She'd arrived to find the door locked, a dead sex worker sprawled across a bed—throat slashed, blood pooling on the mattress sheet—and Hayes clutching another girl, knife pressed to her neck. The girl's gasps carried through the comms, sharp and panicked, as Hayes muttered incoherently.

Lora's call to Kendricks had gone to voicemail twice before he grunted a response, his voice thick with sleep. "Handle it, Vasquez. I'm coming." That was ten minutes ago. She'd pinged Dr. Chen next, anticipating bodies—living or dead. Chen was en route with her nursing assistant, Jacqui Melange. It gave Lora some confidence, Jacqui was a no-nonsense woman with strong arms earned from hard work dealing with a litany of complaints and injuries.

Lora's jaw tightened. Kendricks was dragging his feet, as usual—the Pleasure Quarters weren't worth his time.

Footsteps echoed down the corridor. Mia Hawthorn rounded the corner, her coveralls somewhat askew, hair tangled from sleep. Her eyes locked onto Lora, then the door, reading the situation in a heartbeat.

"Where are we up to?" Mia asked, her voice low but urgent.

Lora kept her gaze on the door. "Hayes has lost it. He killed a girl and is holding another. He has a knife to her throat. I'm keeping him talking till Kendricks gets here."

Mia's face hardened. She stepped closer, pressing her ear to the panel. A muffled sob came through, then Hayes' ragged shout: "Just try it bitch!"

"Ronny!" Mia called, her voice cutting through the comms. "It's Dr. Hawthorn. I know what you're feeling—it's happening to me too."

Silence. Lora shot her a sharp glance, but Mia ignored it, her focus absolute.

"Ronny?" Mia tried again.

"Go away!" Hayes' voice screeched, shrill with paranoia. "I'll cut her! I'll do it!"

"You don't need to do that," Mia said, steady now, like she was soothing a wounded animal. "Let the girl go. You can have me instead. A swap. What do you say?"

Lora's hand twitched toward her stun baton. "Hawthorn, what the hell—"

Mia waved her off, eyes never leaving the door.

"You want to die?" Hayes screamed.

"No," Mia said. "I want you to see we're the same. I know what you're going through."

"You don't know shit! You're a liar! I'll cut her, I swear!"

"You think she cheated you," Mia said, her voice softer, probing. "Maybe she did. But I won't cheat you."

"How do you know that?" Hayes' tone wavered, suspicion mingling with confusion. "How do you know she cheated me?"

"Because you're angry," Mia said. "I'd be angry too. Let her go, and I'll take her place. I promise not to cheat you."

"How can I trust you? You'll cheat me too!"

"You're right, I could," Mia admitted, her honesty disarming. "But I'm not a sex worker. I only care about you. I want to help you."

"No one can help me!" Hayes screamed. "Go away!"

"I can help you," Mia insisted. "Let me prove it. If I'm a liar, you can do whatever you want with me."

A hysterical laugh burst through the comms, jagged and unhinged. "You're crazy! I don't believe you!"

"You better believe me, Ronny," Mia said, her voice becoming steel. "I've never broken my word. You know that's true."

Silence again. Lora watched Mia, her rigid posture betraying the stakes. Hayes was searching his memory—Mia could feel it, the Shard's influence twisting his thoughts, just as it clawed at hers.

Footsteps again, two pairs. Dr. Lilien Chen appeared, her faded medical whites stark under the red light. Jacqui Melange trailed her, clutching a trauma kit, her face taught with nerves.

Seeing Mia at the door, Chen's face fell into confusion. "What's going on here?"

Mia didn't turn. "Back off, Chen. You too, Jacqui."

Chen's lips tightened, but she stayed silent, sensing Mia's resolve. Jacqui looked to Chen for guidance. Chen raised her hand and stayed her.

Mia pressed her face to the door. "I'm alone, Ronny. Open it."

"Are you alone?" Hayes' voice was a low growl, testing her.

"I'm alone," Mia said. "Open the door."

A faint click. The door crept open, revealing a dim room thick with the iron tang of blood. Hayes stood in the center, wild-eyed, his knife biting into the girl's throat. Her neck oozed a thin red line, her eyes bulging with terror.

Mia swallowed hard. One wrong twitch of Hayes' hand and it would be arterial.

The dead worker lay spread-eagled naked on the bed, her throat a gaping wound, her lifeless stare fixed on the ceiling.

"I'm coming in now," Mia said, her voice calm despite her racing pulse. She leaned toward Lora, whispering, "Leave it unlatched. Wait for my signal."

Lora nodded, her expression unreadable but steady.

Mia slipped inside, the door easing shut but not locking. "I'm here, Ronny," she said. "I kept my word. Now keep yours. Let her go."

Hayes' gaze darted to the dead girl, his face twisting with guilt and fury. "I didn't mean to hurt her. She lied to me!"

"Ronny, look at me," Mia said, her voice sharp enough to cut through his haze. "Let the girl go."

Hayes backed into a corner, dragging the girl with him, the knife pressing harder. The girl choked on her own whimper.

"Listen to me, Ronny," Mia said, slowing her steps. "You don't have to do this. No one's going to hurt you. Just put the knife down."

Mia raised her hands, palms open, fingers splayed to show no threat, no weapon. Her heart thudded, but she kept her breathing even, and her movements slow and deliberate. This was no act. She was drawing on all her knowledge of defensive behavior, gleaned from her studies of the brain. Hayes' eyes flicked between her and the girl; his pupils were fully dilated, sweat poured off his brow. The Shard's influence was there—she could see it in the erratic twitch of his jaw, the same 40Hz chaos Chen had charted in her neural scans—yet he wasn't completely gone. There were signs she was getting through. His rage was fraying, giving way to something softer, more fractured—confusion, maybe guilt.

"Ronny," Mia said, her voice barely above a whisper, "you're not alone in this. I feel it too. The anger. The noise in your head."

Hayes' grip on the girl tightened, then loosened, his knife hovering a fraction lower. The girl's breath stifled, her body trembling under his arm, but she didn't dare move. Her neck bled steadily now, the thin red line widening under the blade's pressure. Mia took a half-step forward, her boots scuffing softly on the blood-slick floor. The room seemed to shrink, the red light from the corridor filtering through the cracked door, casting Hayes' face in a feverish glow.

"You don't have to hurt her," Mia continued, locking eyes with him. "You're stronger than this. Stronger than what's pulling at you. Stronger than what broke you."

His knife dipped further, the blade catching the light as it angled. Hayes' lips parted, a choked sound escaping—not a word, but a fracture in his resolve. His shoulders slumped slightly, the weight of his own actions pressing down, the Shard's chaos clashing with a flicker of the man he'd been.

The girl sensed it too. Her eyes darted to Mia, pleading silently. Mia nodded, almost imperceptibly, holding Hayes' gaze, her own fear buried deep beneath her focus.

In that fragile moment, the girl acted. She twisted sharply, her elbow jabbing into Hayes' ribs, breaking his hold. She stumbled forward, hands clawing for the door, her scream raw and piercing: "Let me out! Let me out!" Her fists pounded the panel, the sound a frantic drumbeat that shattered the room's tense hush.

Hayes recoiled, clutching his ears, his knife swinging wildly in the air, no longer aimed. "Shut up!" he roared, his voice splintering with panic. "Shut up!"

Mia edged closer, her voice low. "Ronny, it's just us now. You and me."

The door burst open. Martin Kendricks stormed in, his pistol raised. There was an instant when time itself seemed to freeze, then a volley of shots tore through the room like a system fault, jarring and unnatural. Three rounds slammed into Hayes' chest. Blood bloomed across his shirt, and he crumpled, a stunned expression froze on his face as he hit the floor.

Mia whirled on Kendricks, fury overtaking her. "What the hell are you doing? I had it under control!"

"You're out of line, Hawthorn!" Kendricks snapped, his eyes cold, still half-lidded with sleep. "Get out before I lock you up for sedition!"

Mia stared, shock replacing rage. Sedition?

Lora stepped forward, her voice quiet but firm. "Go, Mia. I'll handle this."

Mia's legs moved before her mind caught up. She stumbled

into the corridor, leaning against the wall, her breath ragged. The surviving girl's sobs echoed from inside, mingling with Chen's clipped orders as Jacqui knelt and attended her. She applied a gauze to the girl's neck while Chen checked Hayes' pulse. Moments later, Chen looked at Lora and shook her head.

Chen got up and left the room, her face a mask of exasperation. "You were reckless, Mia. Risking your life like that—what were you thinking?"

"I wasn't reckless," Mia shot back, her voice breaking. "I was doing the right thing! Kendricks broke the law!"

Chen sighed, her pragmatism a wall Mia couldn't breach. "Come," she said softly, guiding Mia away. "Forget about it."

Mia allowed herself to be guided away, tears stinging her eyes, Chen stealing glances at her as they walked. The corridor stretched endlessly, the red lights flickering like a failing pulse. Behind them, the Pleasure Quarters fell silent, but Mia's mind wrenched. Hayes' death wasn't an accident! The Shard was slowly poisoning the GenStar's soul. And now Kendricks' threat hung over her, a noose tightening with paranoid certainty.

They came to a bulkhead with an air seal. Chen went through first, Mia followed.

They were out of the Pleasure Quarters now and the light was harsher here. A small section of no man's land before the Common Dome. Mia wiped her eyes, blinking. For a moment the whole thing felt like a dream, like she has just emerged from a nightmare.

"Go back to bed," Chen told her. "We'll debrief in the morning. After the autopsies."

Mia looked at Chen for a moment, gathering her thoughts. "No," she said, shaking her head. "This isn't over. Not by a long shot."

Chen put a hand on Mia's arm. "Let it go Mia."

"No," Mia said, her voice low, resolute. "We need to visit Cory Dunbar. If this is a contagion, we need to stop it now."

– 14 –
DESTINY OF THE STONE

The corridor beyond the Common Dome was a maze of steel and polygrout, its overhead lights casting a sterile glare that stung Mia's eyes after the Pleasure Quarters' feverish red. She walked beside Dr. Lilien Chen, their steps echoing in the quiet, a metronome of calm that seemed strangely at odds with the vision of Hayes' form crumpled in the room back there, his neural spikes, charted just three days earlier, now feeling like an abstraction. But they weren't. Cory's were still attached to a body—a living body. The thought snapped her back to the moment, and gave her an added impetus to hurry.

Chen tapped her comms panel, her voice clipped as she hailed Mining Control. "This is Chen. Is Cory Dunbar on shift?"

Static crackled, then a tired voice replied. "Negative, Doc. He's off shift. Been off since yesterday."

"Copy," Chen said, glancing at Mia. She pulled her data tablet from her coat, fingers swiping to call up a map of the residential sector. A grid flickered to life, pinpointing their target. "His quarters—Residential Tier Two. This way."

Mia nodded, her jaw tight. She just hoped they weren't catching this too late.

The residential sector loomed ahead, a warren of sealed doors marked with identifiers in faded stencils. The air here was warmer, heavy with the hum of recycled oxygen, but it did little to ease the chill in Mia's spine. She'd felt her own fears sharpen since the Shard's discovery—doubts about her identity, a need to protect the crew—and was sure now it was more than stress. If Cory's scans meant what she feared, she could have another Hayes on her hands.

Chen slowed, her tablet's glow dimming as she confirmed their position. They stopped at a door labeled R-Tier2-819: Dunbar. She pressed the chime, its soft ping jarring in the silence. Mia and Chen waited in stiff silence. Momentarily, the door hissed open, revealing a slight woman with tired eyes and a braid coiled tightly at her nape—Mimi Dunbar.

"Dr. Chen?" Mimi's voice was soft, wary, her gaze flicking to Mia. "What's this about?"

"Is Cory in?" Mia asked, keeping her tone neutral despite the pulse in her throat.

Mimi hesitated, glancing back into the dimly lit quarters. "He's sleeping. Been out since his shift ended last night."

Chen's eyes narrowed slightly, but she stayed silent, letting Mia take the lead. Mia stepped closer, softening her voice. "Mimi, has Cory been acting … strange lately?"

Mimi's lips pressed into a thin line, her fingers tightening on the doorframe. "Strange how?"

"Different," Mia said carefully. "Mood swings, maybe. Or new beliefs."

Mimi's gaze darted away, her fingers worrying the seam of her sleeve. "I don't know what you want," she said, the emotion in her voice belying her true feelings, "but please, just go away."

The unexpected denial impelled Mia to press harder. "I understand how difficult this must be for you, for all of us—"

"—You have no idea." Mimi said, reaching for the door control.

Mia thrust her hand out and stayed her. She knew it was a radical thing to do, to impose on the freedom of a crew member. It went against all protocols. There was a precarious moment when it looked like Mimi would resort to aggression.

In that moment, Chen found her voice and said, "Hayes just killed a woman."

Mimi froze.

Mia nodded. "It's true. We just came from there. Please, we need to know if Cory's all right."

Mimi's shoulders sagged, as if the request had cracked a dam she'd been holding back. "I don't know what to say …"

"Take your time," Chen said.

Mimi stepped half out into the hallway and lowered her voice. "He's changed. I can tell you that."

"Change how?" Mia asked.

"He just joined Advisor Chandra's group. Calls it the 'Destiny of the Stone.' That's completely out of character."

"Before, he was ..." Mia left the question hanging, hoping Mimi would fill it in.

"Before, he used to believe in the Continuance," Mimi said, her voice dropping to a bitter note. "Now it's all about that thing."

"The Shard?" Chen said.

"Yeah, the Shard."

Mia's brain lurched. This is exactly what she feared. But not in the way she expected. Hayes had turned paranoid, whereas Cory seemed to have become deeply superstitious. Apart from sharing a sudden switch in their cognition, the direction they took was completely different. Both were typical rock rats. As far as she knew, that sort usually only paid lip service to the Continuance out of respect for their families. Down in the mines, it would have been the last thing on their minds. She wondered what role Prya had played in this. Did Cory attend her sermon at the Common Dome? She would have recognized his shock of carrot orange hair. She couldn't say one way or the other, and she didn't want to guess. "Has he ... been violent?" she asked evenly, trying not to sound like she was preempting the question.

Mimi's eyes flicked to a corner of the room, where a small toy ship lay abandoned on the floor. "Only if you disagree with him," she said quietly. "He gets this look, like he's not hearing you. It's just the Stone."

Chen stepped forward, her tone clinical but not unkind. "Don't provoke him, Mimi. Agree, be supportive. Keep things calm."

Mimi nodded, but her hands trembled. "I'm trying. I am. But I'm worried about Jacob." She glanced toward a narrow bunk in the shadows, where a boy's silhouette curled under a blanket. "Cory's filling his head with nonsense about the Stone. Telling him it's our destiny, that it's everything. Jacob's only eight. He doesn't know what to think."

Mia's chest tightened. An eight-year-old caught in Cory's fervor wasn't just a tragedy, it was a warning. Like Hayes' violence, Cory's obsession was spilling over, reshaping those closest to him. She knelt slightly, meeting Mimi's gaze. "If you ever feel you or Jacob are in danger, don't hesitate to contact us. We'll organize a safe place for you to stay. I promise."

Mimi's eyes welled, but she blinked it back, nodding. "Thank you," she whispered. "Both of you."

Chen touched Mimi's arm briefly, a rare gesture from her. "We'll check in soon," she said. "Stay strong."

Mia lingered a moment, her gaze drifting to Jacob's bunk. The boy stirred faintly, his small hand clutching the blanket, oblivious to the mania reshaping his world. Then she turned, following Chen back into the corridor. The door hissed shut behind them, sealing Mimi and her fears inside.

As they walked, the sterile lights buzzed overhead, their hum a faint echo of the GenStar's fraying systems. Mia's mind raced—Cory's cult fervor, Jacob's indoctrination, Hayes' blood on the floor. Something was amplifying the crew's deepest drives, turning faith into zeal, anger into murder. And Kendricks' sedition threat still loomed, a blade waiting to fall.

Chen broke the silence, her voice low. "If Cory's like Hayes, we need his scans updated. Tonight."

Mia nodded, but her eyes were fixed on the corridor's end, where the Common Dome's glow flickered like a failing beacon. Cory's superstition, Hayes' psychosis, her own sharpened fears—they weren't random. Something was spreading, twisting minds across the GenStar. If they didn't find the cause soon, the whole ship could break.

A thought occurred to her.

"You still got the fragment I gave you?"

"Yes, I've been using it to treat patients," Chen said. "It has worked wonders."

Mia stopped. "You're not serious ..."

"I am completely serious. It doesn't cure cuts and other common ailments, but I'm convinced it alters cellular metabolism for the better. I've been planning to run some tests to prove it. This could be the big medical breakthrough we've all been looking for."

"But Lilien, you just saw what it did to Hayes. You seriously can't—"

"Hayes was already unstable before the Shard. What he did tonight was going to happen sooner or later," Chen replied staunch-ly.

"If that's true, what about Cory? What about the riots?"

"It's the ice-mining talking Mia. Even you admitted it. The extraordinary stress that everyone is under."

"I don't disagree with that," Mia began, but Chen wasn't listening and went on, her tone becoming increasingly strident.

"A good scientist wouldn't jump to conclusions about an inert stone before ruling out more plausible factors. You of all people should know that."

Mia's mouth opened, then shut again. She couldn't believe what she was hearing. And then it dawned on her. They were all going crazy! But each person thought they were sane in their own beliefs. She could see the madness in Chen's eyes, while at the same time thinking she herself was completely rational. Meanwhile, Chen was thinking the exact same thing about her! Her mind began to race. If that was true, then she had been going about it completely the wrong way.

"Okay," she said slowly, carefully. "Let's say you're right. Let me prove it. Let's go to the lab right now, and conduct some tests—on each other."

Chen paused, her eyes narrowing slightly, a faint smile tugging at her lips as if she welcomed the challenge. "Fine," she said, her voice steady but laced with a quiet intensity. "Let's do it. But don't expect me to indulge your paranoia, Mia. I'm after answers, not conspiracies."

"I couldn't agree more," Mia responded, matching Chen's tone, "And that's why we're going to get Charlie to conduct the tests on us. So neither of us can accuse the other of cognitive bias."

Chen blinked, surprised. Then nodded. "That's acceptable."

"Good then," Mia said, refraining from saying anything further in case Chen decided to change her mind.

They walked on, toward the lab. Silent. Focused.

But each carried the same unspoken fear:

What if the truth was already behind them?

And what if Charlie—their mirror—was the only one left not human enough to lie, and they were both too far gone to hear the truth?

— 15 —
TEST

"So how do you want to do this?" Chen asked as they entered the lab.

"We'll keep it simple," Mia said. "Take turns. Doesn't matter who goes first."

"I'll go first then," Chen said.

"Okay. Take a seat."

Mia placed the neural cap over Chen's scalp and switched on the scanner. She expected Charlie any moment now. He didn't disappoint, arriving just as she finalized the calibration parameters—his movements brisk, almost eager.

"What's my role in this experiment?" he asked, looking between the two scientists.

"We believe there might be neural contamination affecting both Dr. Chen and myself—"

"Speak for yourself," Chen interrupted sharply.

"—and we need an impartial third party to conduct the questioning," Mia continued smoothly. "Someone who can stay objective."

Charlie nodded. "I've got your baseline scans in memory. I'll monitor for any deviations."

"Perfect," Mia said, stepping away from Chen.

Charlie positioned himself where he could view the neural activity display without letting Chen see it. "Ready to begin, Dr. Chen?"

"I was ready before you got here," Chen muttered, staring straight ahead.

After a moment of calibration, Charlie began. "Please tell me what you recall about the Blowout of 3991."

Chen's fingers tightened almost imperceptibly on the armrests. "Seriously? That again? Why not just ask if I hate my job while you're at it?"

"The question stands, Dr. Chen," Charlie replied, his tone firming slightly.

"Fine. Seventeen casualties. A preventable disaster." Her voice was clipped, clinical. "Happy?"

Mia watched Charlie for any reaction. There was none.

Charlie adjusted the scanner settings. "Among those casualties was Lydia Nikolaidis. Do you remember her specifically?"

"Cut the crap, Charlie. What are you fishing for?"

"Just trying to assess your recollection," Charlie replied. "Lydia's death was particularly unfortunate."

"Unfortunate doesn't even begin to cover it," Chen grated. "Her lungs were crushed to pulp. Even with life support, we barely kept her alive."

"And Rheims Duvall?"

"Similar. Thoracic injuries. Shattered pelvis."

"But that wasn't what killed him, was it?"

Chen's neural patterns spiked for just a second. "You want me to say it wasn't."

Charlie leaned in slightly. "I've reviewed the files. The autopsy documentation had ... irregularities. Missing timestamps, incomplete toxicology screens. Not your usual level of thoroughness. Am I right?"

Mia stepped forward. "Charlie, that's enough."

"I need to press this," Charlie said quietly. "For accurate results."

Mia's mouth tightened, but she stepped back.

"We were overwhelmed," Chen said defensively. "Working around the clock. Mistakes happened."

"These weren't mistakes made in the moment," Charlie countered. "The reports were filed five days later. You had time to get them right. And during that same period, I found requisitions from you—unusual reagents from the Chem Lab."

Chen's face darkened. "Are you accusing me of something?"

She reached for the neural cap.

"—Wait," Mia interjected, eyes locked on the neural display. "We need to finish."

"For what?" Chen snapped. "So your pet bucket of bolts can prove he's better than a human?"

"So we can know what's happening to us," Mia said firmly. "It's my turn next, Chen. And trust me—it won't be any easier for me."

Chen hesitated, then slowly settled back. "I can't believe I'm doing this."

"The injuries documented should have been survivable," Charlie continued. "Statistically speaking."

"Statistics don't factor in the chaos of real-world trauma," Chen retorted, "You weren't there."

"I'm not questioning your medical judgment. I'm questioning the overlap with your experimental work at the time."

"What exactly are you implying?"

"I'm not implying anything," Charlie said evenly. "I'm asking directly: were you testing something that could've affected mortality rates?"

"This is absurd." Chen said, yanking the neural cap off her head and tossing it onto the bench. "I'm not sitting here to be accused of malpractice—or worse."

Mia picked the cap up.

Chen stalked to the far end of the room, furious.

Mia stood there for a moment, assessing Chen. This was not the controlled, efficient Chen she knew. Her breathing was sped up, her hands clenched and unclenched, and there was a distinct sheen of sweat across her brow. Chen was staring back at her, but Mia sensed she wasn't looking at her so much as through her.

Charlie broke the spell. "I'm sorry I had to put you through that Dr. Chen. I can assure you, it wasn't my intention to cast aspersions. You can rest assured, the data will remain confidential."

Chen gave Charlie a cold grin. "But you got exactly what you wanted, didn't you?"

"Again, don't confuse what I was asked to do and what was needed," Charlie said. "Why don't we hit the pause button for a moment and turn our attention to Dr. Hawthorn. You might learn that you are not alone."

Charlie's soft but authoritative tone represented the best bedside manner that any doctor could have asked for.

After a long pause, Chen said, "Fine. But don't think I won't forget this incident."

"Is that a threat?" Charlie asked.

"You can take it whichever way you like," Chen said, dismissively.

Mia said to Chen, "I'm ready to go, if you want to help set me up ..."

Chen huffed noisily and came over. She picked up the neural cap and placed it over Mia's scalp, somewhat roughly and without making eye contact.

Charlie began the calibration. Satisfied, he turned his attention to Mia, and in the same low, even tone he had used with Chen, he said, "Close your eyes, Mia. I want you to remember the reactor incident. Picture the control room. Your father."

Mia had expected this. Even still, her breathing rate increased ever so slightly. "I can see it," she said. "What would you like to know?"

"Can you see you father?"

"Yes, he's monitoring the reactor pressure vessel. Following protocol."

"Go on."

"There's a loud bang. A seam cracks. A bolt pops off like a champagne cork and flies across the room—it hits Rivel in the eye, blinds her."

"They're the only two in the room?"

"Yes. Dad's on the floor. Rivel's up on the gantry. Control is watching through the observation window."

"Then what?"

"The pressure vessel gushes superheated fluid. It floods the floor fast. Dad's up to his knees in it."

"He doesn't run ..."

"Control's shouting over the PA. The noise—steam hissing, pipes shrieking—he can't hear them. But he sees the hand signals. They can't shut the valve from the panel. It has to be manual."

Mia's chest rose rapidly. Sweat beaded on her forehead.

"You're doing fine," Charlie said. "Keep going."

"He wades across the room. Finds the valve. Turns it off. By the time he's out, the flesh is stripped from his calves."

"But he saved the day."

"I can still see him—on the ventilator in the infirmary. In a coma. His face pale, hair matted, legs wrapped in bandages ... oozing."

"You can rest now."

Mia opened her eyes, drained.

Watching Mia, Chen's reaction had softened. She seemed to have regained her composure, business-like, efficient.

Mia pulled off her cap and turned her attention to the analysis of the data. She went over to the lab computer and started collating the results.

The three of them leaned in toward the monitor.

"There. See that?" Mia pointed to a spike in Chen's graph—right at the moment Charlie pressed her about the autopsy report.

"I see it," Chen said, begrudgingly.

Mia pulled up her own graph beside it.

"You too," Chen said, noticing a similar spike when Mia relived her father's trauma.

"Yes ..." Mia said, not surprised.

"But that's not the most interesting part," Charlie said, pointing to a divergence between the deeper cortical regions and the higher processing areas. "Run a correlation on that, will you?"

Mia punched in a few parameters, pulled a logarithmic view of the region.

"Holy darn!" she said. "Our limbic systems are functioning normally, but the cortices are overclocking."

"That shouldn't even be possible," Chen muttered, rubbing her eyes. "That's not how the brain works."

"Seems like whatever's affecting you," Charlie said, "hasn't taken over your emotional centers yet."

"But for how long?" Mia asked, locking eyes with Chen.

"I'm not denying it anymore," Chen said. "Our brains are changing. But I still think I'm rational."

"I would've said the same until I saw these results," Mia said. "What makes you think you're immune and I'm not?"

"You're interpreting the data to fit your fears, Mia. I don't see it that way. If anything, maybe our cortical regions are suppressing our limbic responses."

"She's got a point," Charlie said. "But it still doesn't answer the question—what if the limbic system's next?"

"We'll cross that bridge when we get to it," Chen said. "As far as I'm concerned, these tests confirmed what I already knew." She clapped her hands, like ending a lecture. "If you'll excuse me, I've got to get back to studying the Shard's effects on my patients. I'm already behind."

With that, she swept out of the room.

Charlie and Mia stood there, staring at each other.

After a while, Mia said, "She's in denial."

"You might be right," Charlie replied. "But the question is—what are you going to do about it?"

"I'm starting to think Kendricks was right."

"About what?"

"Getting rid of it. Ejecting it into space."

Charlie's expression shifted—grave, almost pained. Mia had never seen that look before.

"What is it, Charlie? What's bothering you?"

"I'm worried about your safety. You won't be able to move the object alone. You'll need help."

"Are you saying you'll help me?"

Charlie raised and lowered his hands, like he didn't have a choice.

"Then it's our secret," Mia whispered.

"Yes."

"Wait for my signal," she said. "There's something I need to do first."

— 16 —

ROOTS

The Agriculture Sector breathed from the constant sound of fans pushing air through its long tunnels. Mia stepped through the entrance archway and immediately felt the change. The air was thicker, heavier with moisture, and carried a sweet, cloying scent. A stark contrast to the dry, stale odor that lingered in the other sectors. As she moved along a row of hydroponics, she saw condensation dripping from pipes. It was the closest thing to rain she had ever experienced in her confined, artificial life. She passed a jumble of wires and power transformers stacked on a trolley. She wondered if it was Callum's handiwork, since maintaining the grow lights was one of his main responsibilities.

Her boots stuck slightly to the algae-slick floor as she followed the violet glow of the algae tanks. It was still pre-dawn, but that didn't stop her. The person she was looking for worked the graveyard shift. There, perched atop a stack of empty crates, sat Sonia Hawthorn, balanced like a gargoyle, legs crossed, a jar of amber liquid resting against her knee. At first she didn't seem to recognize Mia, but then she slowly registered the person approaching her.

"Well, hello stranger," she said in a voice that was roughened by harsh drink. "Come to gawk at the miracle of photosynthesis, or looking for something to pinch?"

Mia stopped short, keeping her distance. Her shirt under her coveralls was already damp.

"You always hated this sector," Sonia went on. "Couldn't stand the slow pace of life here."

"I was a kid," Mia replied defensively.

Sonia screwed her eyes shut then snapped them open with conscious effort. There was more focus in them now. "You're still my Mia," she said with conviction. "Somewhere under all those barnacles encrusting your soul."

Mia hesitated. Her mother's observation grated against her sense of self, but she decided to ignore it. "I saw Tiber yesterday," she said innocently as possible.

"How's he doing?"

"He's okay, I guess. We're all a little worse for wear."

Sonia snorted. "Go tell that to Dr. Chen. Good luck squeezing an extra vitamin D ration out of her."

"Can we talk?"

The sarcasm slid off Sonia's face. Her gaze landed heavy on her daughter. "You haven't been around in who knows how long. What could we possibly have in common anymore?"

"Dad," Mia said simply.

Sonia recoiled, the memory striking like a sudden shift of gravity. She gripped the jar tightly in both hands, as if it was the only thing that was solid and steady in her world.

"I saw him today," Mia said, leaving her mother no option but to respond.

"You *saw* him?"

"In my mind. Clear as the day it happened."

Sonia gave Mia a double take. "Not you too," she groaned.

"What do you mean?"

"I've heard the stories ... that strange object they found down the mines. Prya's been parading it around like a sacred relic. Don't tell me you've gotten caught up in her nonsense."

"No," Mia shook her head firmly.

"Well, that's a relief. For a minute there I thought you'd gone to the fairies."

"You know me better than that, Mom."

"Do I?"

"I wanted to hear your version of the story, that's all. After all these years, I thought I'd forgotten it."

Sonia shook her head slowly. "You don't forget. Not really." She held up the jar of homebrew. "Not even Dr. Liquor here can scrub that from the memory banks."

Mia watched her mother take another gulp. "Is it true? That he saved the whole ship?"

"A hero if there ever was one. But don't tell him I said that," she added with a hoarse laugh.

"So it's true."

Sonia blinked slowly, then rose from her perch. She moved across the tunnel with the careful steps of someone used to walking while buzzed. She stopped before a rusted cabinet. She didn't open it. She stood there for a while, seemingly debating with herself.

Mia watched her without saying anything.

Without turning around, Sonia said, "I suppose now's as good a time as any."

"For what? Mia asked.

Sonia didn't answer. She opened one of the cabinet's doors, its hinges protesting with a creak. She rummaged for a bit, then pulled out a small plastic case.

Mia watched with growing curiosity.

Sonia turned around and held the case out to Mia.

"You want me to have this?" Mia asked.

Sonia didn't answer. She simply pushed the case into Mia's hands.

Mia studied it for a moment, found a small latch on one side. She flipped it and opened the lid.

It was a medal, deep bronze, stamped with the Arkonauts'

multi-generational symbol—three figures holding hands, each connected to the next, fading into an infinite horizon. Around the rim were the words: *Firmate Catena Maiorum*. Strengthen the Chain of the Ancestors.

Sonia said, "I don't know why I kept it all these years. But there you have it."

"You're giving it to me?" Mia asked incredulously.

"Like I said, now's as good a time as any."

"Thank you," Mia said, hugging her mum. It was brief. Sonia didn't let it last.

Mia snapped the case shut and dropped it into her pocket.

Sonia said, "True to form, his last words were: 'Have they safed the reactor?'"

Mia flinched. "I didn't know he spoke before he died."

"That's because the official inquest buried it. I had to dig it out of the eyewitness reports. Tiber Solis, bless his heart."

Mia felt her eyes sting. She wiped them with her sleeve. "I was scared I'd forget," she admitted.

Sonia rested a hand on her shoulder. "Not possible. I wouldn't let you."

"But you wanted to forget."

"No. I just wanted the pain to stop."

"Maybe the two aren't so easy to separate," Mia offered.

Sonia gave a shallow laugh. "How much do I owe you for that little nugget of wisdom, Dr. Hawthorn?"

"It's on the house. Perks of nepotism."

"So you came for my memories. What about me?" Sonia asked. She picked up her jar again and took a hefty swig.

"I'm just glad you're not showing symptoms like the others," Mia said, relieved.

"You mean I'm not talking to the eggplants, like the others?"

Mia laughed. "As long as they're not talking back to you."

"Well, that seems to be the thing, doesn't it?" Sonia said. "It seems like that stone is driving everyone nuts. I don't know ... feels like it's amplifying all our little quirks, turning them into obsessions."

Mia looked at her mother in wonder. "Like an amplifier. You're right."

"Which means it must have a field. Get too close and you fall into its basin of attraction."

"Field ..." Mia echoed.

"I'm no physicist," Sonia said. "I'm a botanist. I think in terms of bioelectric fields. That kind of thing? I don't even know what it is. All I know is people can't stop talking about it. Like they've completely forgotten the Landing Craft."

"A bioelectric field ..." Mia murmured. "You might be onto something there."

Sonia sloshed her drink in the jar, watching it swirl. "Don't listen to an old drunk like me. What do I know?"

"This bioelectric field ... do you think it could be operating in the gamma range? Forty to a hundred hertz?"

Sonia exhaled through her nose. "Plants have been known to

respond to all kinds of frequencies. What's a brain? Just a big mel-on. Right?"

"I suppose," Mia said, not wanting to disagree.

"Have you ever heard of the metamorphic organizing prin-ciple?" Sonia asked. "A seed becomes a tree. An egg becomes a person. Sure, genes govern development. But that's not the whole story, is it? Something special happens when cells cluster and start communicating. A shared memory emerges. It's the mystery of consciousness. What if this thing is amplifying that?"

Mia frowned. "It's possible ... although the symptoms are dif-ferent for each person."

"Well that's your common thread right there," Sonia said off-handedly.

"How so?" Mia asked.

"Each person goes crazy in their own way. But crazy they go."

Mia couldn't help agreeing. "It's as if its activating a pre-exist-ing complex, an unconscious fixation."

Sonia firmly nodded.

"The fixation was always there. The Shard merely triggers it ..." Mia murmured.

"Hey! What does anyone care what I think?" Sonia said, rasp-ing a laugh. "I'm just a crazy old bitch who thinks my plants can hear what I say."

Mia ignored the comment and said, "So how do we stop it?"

"You don't. It's baked into the system."

"Then we're all doomed," Mia said.

Sonia shrugged, not seeming to care. "We were always doomed."

- 17 -

THE SHARD OR US

On the way back from the Agri-Sector, Mia went to the Common Dome restroom, relieved herself, then grabbed a tepid drink from the food dispensary before heading to her private quarters. It was a little past 3 a.m. She didn't know what to expect, but when her door hissed open and the corridor light fell across Devory's sleeping form in her bunk, she felt strangely relieved. She could almost convince herself again that everything had returned to normal. Except that it hadn't.

She quietly stepped inside and closed the door behind her, the room returning to darkness, the only light remaining came from the LED on her touchscreen display beside her bed. She silently went over to the chair next to her bed and undressed. As she did so, she felt the weight of her father's medal in its box inside her pocket. She took it out and put it on a small shelf next to her bed. She was about to climb into bed, when she paused for a moment, watching Devory's faint outline in the near darkness, the bedspread covering him rising and falling in a steady rhythm. As much as she didn't like him taking sleeping pills, she was glad that he had slept through her entire encounter, free from the worry she had gone through.

She lowered herself onto the edge of the mattress. The subtle movement was enough to stir him.

"Mmmm." His eyelids fluttered open. "Mia?"

"Yeah," she said softly. "It's me."

"What time is it?"

"It's still early. Sleep."

"You woke me up."

"Sorry."

He shifted his arm from out under the blanket. His wristwatch cast a ghostly glow across his face. Mia looked into his eyes and saw that they were free from that intense, feverish focus that had gripped him earlier.

"Where were you?"

"Busy."

He yawned and said, "I remember now. You said something about the Shard ..."

"It's fine," she lied.

Devory looked at her, got up on one elbow. His hair was disheveled, his face haggard. "I was so tired when you left, I crashed." He gazed at her for a while. "You've been up the whole time?"

"Yeah."

Devory's features sharpened into alertness. "What's wrong?"

Mia hesitated. As much as she didn't want to say anything, she could see by the way Devory was looking at her that he was going to find out one way or another. She decided that he may as well hear it from her first. Slowly, choosing her words carefully, she said, "There was an incident, but it's under control now."

Devory studied her, his expression hardening. "And?"

"I don't know how to say this ... but Hayes is dead."

"What?"

"I'm sorry."

"What? No! I don't believe it."

"It's true. He killed a sex worker, then took a hostage. I tried to intervene, but Kendricks killed him."

Devory recoiled. He became agitated. "I don't believe it. Who did he kill?"

"I'm pretty sure it was Cynthia Watts."

"What do you mean 'pretty sure'?"

"I didn't hang around; Kendricks was on a rampage. Lilien and I got out of there as fast as we could."

"Lilien?"

"Yeah, she was there as well."

Devory pulled the blanket back and swung his legs out. He ran his hands through his hair, releasing a cloud of dust into the air.

"I need to find out what happened. This is going to hit the team hard."

Mia grabbed him and stayed him. "Dev—listen."

"Listen to what?"

"It's too late to do anything about it now. The other men on shift won't know about it, and the rest are asleep."

"Doesn't matter. I've got to talk to Kendricks and sort this out." He scooped up his coveralls from the floor.

"Dev—"

"What?"

"There's something else we need to talk about."

The urgency in Mia's voice gave him pause. "What? What is it?"

"It's about us."

"What about us?"

"The Shard ... it's changing us."

Devory looked at Mia like she had lost her mind. "Not now Mia. Get some sleep." He got up and slipped on his coveralls.

"Don't you want to know about Cory?"

That got his attention.

"What about Cory?"

"Lilien and I went to his private quarters. He was asleep, but his wife came out and spoke to us. Mimi—"

"I know who Mimi is."

"She said that Cory had become obsessed with geometric shapes. He's fallen into Prya's magic cult."

The semi-annoyed look on Devory's face abruptly changed

into a wry smirk.

"What? What is it?"

"He's been like that since we pulled the Shard. We don't take it seriously."

"Just like you didn't take Hayes seriously?"

Devory's expression darkened again. "Hayes was a loner. He didn't have any family. He spent all his time in the Pleasure Quarters. What do you want me to say?"

"I want you to acknowledge that the Shard is changing us."

Devory gave a dismissive wave of his hand. "I think you're over-dramatizing things. Okay, so Hayes went off the handle. But before you start attributing it to the Shard, I think you ought to consider the hours he was working. He was a sucker for uppers. I'm sure that had something to do with it."

"Possibly ..."

"You see? It has a perfectly rational explanation."

"Maybe."

"Listen," Devory said, placing his hands on Mia's shoulders and gazing directly into her eyes. "I understand your interest in the Shard. It's perfectly natural. We're all interested in it. Speaking of which—"

"What?"

"That fragment I gave you. What did you do with it?"

Mia tried to keep a straight face, but she faltered. "I gave it to Lilien."

"You what?"

"After studying it. She demanded her share."

Devory rubbed his eyes with disappointment. "I thought we agreed it would be our secret?"

"We did, but she was demanding."

"So you caved in?"

"What choice did I have? She would have reported me to Kendricks."

Devory pulled at his hair.

"In any case," Mia said, "I discovered that it emits some kind of field—I think."

"A field? You think?"

"Yes. It seems to have some sort of photoelectric effect at the molecular level."

Devory just stared at her.

"And it's possible that's how it affects neurons."

"Neurons?"

Mia forcefully gesticulated with her hands. "Yes! That's why I think its changing us."

"You want me to take you seriously?"

"Yes!"

"But Mia, can't you see?

"See what?"

"The Shard is the most important discovery in history. It doesn't matter whether it's affecting us or not. We're the first humans to lay eyes on it. Whatever sacrifice is necessary, we have to make it. Do you understand?"

Mia realized her words were just bouncing back. A feeling of defeat washed over her. She nodded vaguely.

"You don't seem to be with the program," Devory said, standing back.

For a moment, Mia thought about escalating the argument. But she changed her mind. "I'm worried," she said, realizing it sounded flat.

"Don't worry," he said, zipping his coveralls up. "The most important thing right now is to sort this Hayes business out. I have to go and rally the team." He bent forward and gave her a peck on the forehead.

She watched him move to the door. He opened it, sucked in a deep breath of the outside air.

"Dev?"

He turned and looked at her. "What?"

His face was fully illuminated by the corridor light now, and he seemed to look different, more assured, more motivated than the old Devory. She thought about telling him her plans to ditch the Shard, but she thought the better of it. He wouldn't agree. She knew it by the way his eyes had returned to that hypnotic stare, the intensity showing in the darkness of his pupils. "Be careful," she heard herself saying.

Devory nodded, his silhouette sharp against the corridor's glare. "I will," he said, but the fervent edge in his voice betrayed him—Mia knew that as soon as he had talked to his men about the Hayes's incident, he would return to searching for pieces of the Shard. She could see it in the way he leaned forward, as if he was about to set off on an epic expedition.

The door hissed shut, sealing Mia in darkness save for the LED's faint ambiance. She sat motionless on the bunk, the warmth of Devory's presence lingering like an afterimage.

She'd almost told him: I'm going to eject the Shard. She shuddered at the thought of what she was planning. It was crazy, she knew it. A parade of names passed through her mind, but in the end, she knew the only person who might understand was Charlie. She could trust him. And she knew that he would trust her.

She rose, the polygrout floor cold beneath her feet, and put her coveralls back on. Tired as she was, she needed to act now. Slowly, with a supreme effort of will, she moved to the door and opened it. She looked up and down the corridor. Devory was gone.

She slipped back outside.

— 18 —
TROJAN HORSE

Mia's heart thudded as she approached the Archives, her boots whispering against GenStar's worn corridor. For countless visits, she'd come to seek Tiber's wisdom, his help for her Drift research. How ironic, she thought. All these years, the concept of Drift was just that—a concept. It could only be written about in the past tense. Minds and cultures changing slower than glacial ice. Now all of that was ancient history. The Shard made a mockery of time. It was evolution on steroids. She was past any sense of doubt. She could see it happening before her very eyes.

She stepped into the Archives, where the air always seemed cooler, quieter, denser with memory. Flickering lights cast uneven shadows across the vast stacks of digital memories and book cases, their solidity a symbol of reassurance. Charlie was at his usual post, head angled to one side in calm attention, amber eyes glinting. Tiber lay curled up in a makeshift cot just beyond, sleeping deeply, his snoring audible over the background hum of the air-conditioning.

Mia tiptoed past him.

"No usual hello?" he asked from somewhere underneath his blanket.

"Sorry, I thought you were asleep."

"Nah, I was just dozing."

"You can keep dozing if you want," Mia said gently.

That made him throw his blanket back.

She stepped closer, the familiar glow of the console between them. "I need Charlie."

Tiber tilted his head. "I suspected as much. May I ask what for?"

She met his gaze, steady. "To eject the Shard."

The words hung in the air, heavy with foreboding. Tiber didn't flinch. He only nodded slowly, his expression unreadable.

"I thought it might come to this," he said, getting up. "And I thought ... if anyone was going to do it, it would be you."

Mia let out a breath she hadn't realized she'd been holding. "You always seem to be one step ahead of me."

"Not always," he said. "Just when it matters."

He went over to his station and leaned forward, tapping a file open with a flick of his finger. "You ever hear of the Trojan

Horse?"

She shook her head. "No. Should I have?"

Tiber chuckled softly. "Old Earth myth. The Greeks were at war with the city of Troy. It was a long, brutal siege. They couldn't break the walls. So one day, they pretended to retreat. Left behind a gift—a giant wooden horse. The Trojans debated whether to bring it inside. One priest warned them not to. Said it could be a trick."

"Was it?"

"It was," he said. "That night, a group of soldiers hidden inside the horse climbed out and opened the gates. The Greek army poured in and sacked the city."

Mia frowned. "So ... the Shard is the horse?"

He shook his head. "No. The Shard's already inside the gates. You're the priest."

"Oh." She let the thought sit. "So if I'm the priest, what does that make Prya Chandra?"

"A fanatic," Tiber said, voice firm.

Charlie turned then, speaking for the first time. "I can requisition a tug-trolley from Engineering. Whip up a manifest that will say it's for transporting ultra-dense water filtration units. Rivel signs off on those without checking. No one will question it."

Mia blinked at him. "You already worked this out?"

Charlie nodded. "I've run simulations. Optimal time is the graveyard shift. Minimum crew density along the transit corridor. If we cover the Shard with something, we should be able to maneuver it to the Master Airlock without being detected."

"And eject it?"

"Yes."

Tiber rose slowly from his chair. "You'll be going against everything Carson, Chen, even Devory are clinging to. You may be condemned for doing this. You still want to do it?"

"I'm used to being misunderstood," Mia said, glancing toward Charlie. "And I won't be alone."

"No," Tiber said, "and paradoxically, Kendricks might even be with you on this one, although I'm sure he'll cook up a legal excuse to hang you."

"Thanks for the comforting thought," Mia said.

There was an awkward moment where neither Mia nor Charlie were prepared to leave.

Tiber took the initiative and guided them to the door. On the way, he touched Mia's arm—not to stop her, just a tether of memory. "Good luck, Mia. Be careful."

She gave him a faint smile. "I'll try."

"And Charlie," Tiber added, "you've come a long way from being a scribe."

Charlie blinked slowly. "I am adapting."

Tiber watched them go, alone again in the chamber of half-forgotten truths. As the door sealed behind them, he murmured to the empty stacks:

"Some gifts are never what they seem."

— 19 —

SECURITY

The rusted pipes and electrical conduits that lined the service tunnel snaked along the walls like exposed arteries. Mia followed them based on their color coding, the easiest method to find your way around in the rabbit warren of tunnels that crisscrossed the GenStar's internal habitat.

Charlie had slipped ahead to Engineering to requisition the tug-trolley, his amber eyes steady with calculated purpose.

As Mia hurried along, a mounting dread gnawed at her for what she was about to do. What if getting close to the primary Shard would send her mad like it had done to Hayes and Dunbar and, as much as she feared to admit it, Devory? If there was one consolation, it was that she believed that by getting rid of it she might bring Devory back. Bring everyone back. That alone should make it worth it.

Her comms unit on her lapel buzzed, sharp and jarring. "Hawthorn, my office. Now."

Kendricks.

Mia's first reaction was disbelief. But there was something in Kendricks' tone, like an iron chain grinding with menace. "What—now?" she said, hoping she was mistaken.

"Now!" came the reply.

Mia's stomach knotted. She had no choice but to obey. She took the next branch and turned toward the Security Wing, its narrower corridors colder, lined with surveillance nodes blinking like watchful eyes. Kendricks summoning her at this hour, after Devory's departure—surely this was no coincidence.

The Security Wing was a fortress of stark efficiency: a steel desk buried under data slates, walls scarred from years of neglect, the air thick with recycled sweat and tension. Kendricks stood, his broad frame filling the space, eyes hard as flint. "I just spoke to Murchison," he said, his voice low and accusing. "What exactly did you tell him?"

Mia squared her shoulders, meeting his glare. "I told him the truth."

Kendricks stepped closer, knuckles whitening as he gripped his side baton. "Truth? He blames me for Hayes' death. Says I was reckless, gunning him down like a dog. He's reporting me to Carson, demanding I explain myself."

"That's your problem," Mia said, her voice steady despite the pulse hammering at her temple. Devory's betrayal stung—she'd warned him about the Shard, not Kendricks' actions.

"Your problem too," Kendricks shot back. "I'll tell Carson you rushed in, playing the reckless hero, forcing my hand to save your ass. You think he'll believe you over me? You, with your Shard fragment and lab conspiracies?" His finger jabbed the air, inches from her chest. "Carson trusts me to keep order, not some scientist chasing ghosts."

Mia's mind flashed to Hayes' paranoia, his knife at a hostage's throat. "You shot him without hesitation. Devory's right to question you."

Kendricks' face darkened. "Question me? You're the one hiding things. I saw the way you and Chen scurried away, tails between your legs. What were you plotting? Another stunt to undermine me?"

"I was doing my duty," Mia snapped, heat rising. "We went to Cory Dunbar's quarters, to make sure he was okay."

This information seemed to catch Kendricks off-guard. "And, was he?"

"If I say no, will you shoot him too?"

Kendricks' face contorted with sarcasm. "What do you take me for? Some kind of animal? Cut the bullshit Hawthorn. Tell me straight. Was he good or not?"

Mia held off answering for a moment, deliberately, to test Kendricks' patience. "Yeah, he was all right. If you can call being obsessed with geometric shapes normal."

Kendricks absorbed that, as if it was some kind of riddle. "I see," he said. "I'll make sure to visit him today. See what he's up to."

"You do that," Mia said, turning to leave.

"Where do you think you're going?" Kendricks called after her.

"To get some sleep," Mia said, stopping.

"We're not finished here until I say we're finished."

Mia slowly turned, walked back, stopping just as their faces came within spitting distance of each other.

Kendricks said, "You're dangerously close to sedition Hawthorn." He indicated with his thumb and index finger, holding them right beneath her nose. "One more misstep, and I'll have you stripped of your rank for insubordination. You got that?"

Mia drilled her gaze directly into his. "Is that all?"

Kendricks' face flushed with barely suppressed rage. But words failed him.

Mia nodded. She turned this time, and didn't look back.

She stepped into the corridor, the door hissing shut behind her like a guillotine. The very moment she was alone, her hand flew to her throat. Her breath came in rapid, shallow gasps, her vision blurring as panic surged in her chest. Not now! She slumped against the cold wall, counting lung-fulls—ten, twenty, forty seconds—until the chaos subsided.

She walked away slowly at first, then gradually picked up pace, walking faster and faster as her resolve sharpened.

As she moved through the tunnel system, Tiber's voice came

back to her: *ripples spread out*. If her mother's bioelectric theory was correct—and it had to be—it was imperative she kept the amount of time spent in proximity to the Shard as short as possible. She mentally envisioned the path from Devory's Lab to the Master Air Lock. The maze of corridors she and Charlie would need to negotiate was daunting. Perhaps if she limited her time to getting it onto the tug-trolley, then kept her distance while Charlie transported it, that would be one way of keeping her exposure to a minimum.

'Exposure'—funny how that word now carried weight. For what it was worth, the moment she and Charlie started lifting it onto the trolley, the clock would start ticking. How long did Devory spend close to it? An hour, two hours, three? It was a scary thought.

The Engineering Bay's entrance glowed ahead, the muffled sounds of machinery reverberating down the corridor. Charlie was waiting outside, his battered titanium alloy frame on standby, amber eyes scanning her approach. "Tug-trolley's secured," he said, voice calm. "Rivel signed the approval, no questions."

— 20 —
MASTER AIRLOCK

The GeoLab's flickering lights cast stark shadows across Devory's fortress of science, its aging equipment silent since the Shard's arrival disrupted his meticulous control.

Charlie stood beside a workstation, his titanium alloy fingers working methodically at the security panel mounted on the wall. "Disabling surveillance cameras," he said, voice pitched low despite the empty lab. "Looping previous footage. Eighteen minutes before the system runs a diagnostic."

Mia nodded, her gaze fixed on the Shard. It hovered in its synthetic net at the center of the room, suspended in defiance of physics—a jagged obsidian teardrop that seemed to pull light inward rather than reflect it. Even now, after everything she knew about its effects, she felt the tug of fascination, a neural whisper urging her to study it, understand it, possess it.

"Focus," Charlie reminded her, noticing her lingering stare.

"Right." Mia shook her head, forcing her attention back to the task. "The tug-trolley?"

Charlie gestured toward the entrance where the heavy industrial trolley waited. Its wheels were individually powered by electric motors. "I just hope the Shard doesn't mess with its electronics," he said guiding it in.

"It's negative matter," Mia said, repeating Devory's assessment with a grimace. "Or something close to it. The quantum field effects alone ..." She stopped herself. The Shard's cognitive pull was subtle, dragging conversations toward theoretical obsession.

"Let's move quickly," Charlie advised, stepping toward the containment controls.

Mia untied the straps from the stanchions that supported the beam the Shard was suspended from.

"On my mark. Three. Two. One."

Charlie maneuvered the trolley beneath the Shard, which paradoxically began to start rolling away as Mia lowered it.

"Put the brakes on Charlie. Hold it steady."

Mia kept lowering the Shard, until ... it refused to lower any further. It hovered over the trolley base, threatening to slip sideways. Charlie stepped forward and stayed it. While he did that, Mia started pulling the net away. It took some fiddling, but eventually, she exposed the Shard fully, held precariously by Charlie, his hands being repelled from actually touching it, but still constraining it from escaping his clutches.

"Quickly," he said, "the tarpaulin."

Mia grabbed the ropes attached to the tarpaulin and one by one tied it down so that the end result was what looked like camel's hump.

Mia released the breath she'd been holding. "Let's move."

Charlie said, "We'll take service tunnel 5-C. It's the most direct route to the Master Airlock."

"It's your call," Mia said.

Charlie grabbed trolley's control yoke. The wheeled platform hummed to life, hovering a few centimeters above the ground. Together they guided it from the lab, Charlie maintaining a steady hand on the controls while Mia checked their surroundings, pulse quickening with each intersection they approached.

The service tunnels were mercifully empty, most of the crew either on duty elsewhere or asleep during the artificial night cycle. Steel pipes and conduits lined the walls, occasionally dripping condensation from decades of minimal maintenance. Their footsteps echoed dully on the polygrout floor as they progressed deeper into the ship's arterial network.

With each step that brought them closer to the airlock, Mia felt her anxiety levels rise, despite her best efforts to keep her emotions under control.

Charlie noticed, as part of his standard operating protocol. "Your stress levels have risen significantly over the last 10 minutes," he said in a neutral tone. "The cause most likely stems from two possible sources. Either you're scared we're going to get caught or, you're having second thoughts. Do you want to talk about it?"

"No," Mia said simply.

"Talking about it can help," Charlie reminded her.

"The answer is still no," Mia said, more firmly this time.

Another minute passed in silence.

Eventually Charlie reached a threshold where his programing forced him to verbalize what he thought was driving Mia's stress: "After careful consideration, I have concluded you're having second thoughts," he said without sounding judgmental, if that were possible in a situation like this.

"I didn't ask for your opinion," Mia said, unaccustomed to the simulant initiating conversation. Perhaps it was the unusual circumstances that prompted it she reasoned.

"I'm sorry, you're right," Charlie said. His eyes briefly flashed orange, signifying self-reproach.

The marched on in silence for another couple of minutes.

Unable to bear the tension that had festered in herself and between them, Mia eventually blurted out: "Don't you think it's a bit too late to be asking that now?"

"Excuse me?" Charlie responded.

"You think I'm having second thoughts," Mia said, reiterating Charlie's earlier comment.

"Oh yes, that," Charlie said. "I was wondering if maybe a part of yourself realizes the gravity of this situation in all its ramifications."

"Stop messing with me Charlie. Just spit it out."

"Well, if you put it that way ... I fully acknowledge the dark side of the Shard, we've seen it with our own eyes. But we've also seen its effects on neural tissue."

"What are you saying Charlie?"

"I'm saying, so far we've only seen the short-term effects. What if the long-term effects are beneficial?"

"Yes, that thought did cross my mind," Mia said, taking up Charlie's challenge, but almost as soon as she said that, she shut it down, saying, "But you forget that not all brains are equal. When you factor that in, the balance falls on the side of negative effects."

"You're right," Charlie intoned. "I was being too idealistic."

"Watch it!" Mia cried.

The tug-trolley scraped against the side of the tunnel wall. Evidently Mia's criticism of Charlie's comment had caused him to lose focus.

"Sorry," Charlie said, steering the trolley away from the wall.

They turned a final corner and at last they arrived at a heavy bulkhead door marked "CAUTION: PRESSURE DIFFEREN-TIAL – AUTHORIZED PERSONNEL ONLY." The Master Airlock lay beyond—a massive chamber designed for major equipment transfer between the ship's interior and the vacuum of space.

"Stand by," Charlie said, interfacing with the access panel. His mechanical fingers moved across the keypad as he bypassed security protocols. "Authorizing emergency maintenance protocol. Override: Engineering Code Alpha Seven."

The bulkhead hissed and began to slide open. Charlie disconnected quickly. "I've bought us maybe a minute at most while the mainframe computer diagnoses a null event. After that, we'll be visible to security."

Mia nodded and helped guide the trolley through the widening aperture. The Master Airlock chamber beyond was cavernous compared to the tunnels—a cylindrical space forty meters across with reinforced walls and a segmented outer door large enough to accommodate industrial equipment. Its rampart was also wide, and could act as a staging area where larger items could be broken down for transport through the narrower tunnels. To accommodate this, it had a crane and jib hoist as well as some shelves for storing items.

"Pressure chamber controls are over there," Mia said, pointing to a reinforced booth beside the massive outer door.

They maneuvered the trolley toward the center of the chamber, its wheels echoing hollowly on the metal floor. The Shard remained suspended beneath the tarpaulin, wobbling slightly from the vibrations transmitted through the trolley.

"I'll initiate the depressurization sequence," Charlie said, moving toward the control booth. "We'll need to secure ourselves before—"

The bulkhead door through which they had entered suddenly hummed back to life. Mia spun around, heart hammering against her ribs.

Charlie froze, his amber eyes widening marginally. "We've been detected."

Mia had no idea how long it would take before someone turned up. She instinctively dropped her hand to the utility knife at her belt. The bulkhead completed its cycle, revealing three figures silhouetted against the tunnel's dim lighting.

Prya stepped forward, her feminine features distorted by the chamber's harsh overhead lights. Her eyes burned with zealous intensity, and behind her stood two acolytes—crew members Mia recognized from Maintenance, their expressions mirroring their leader's fervor.

"Betrayer," Prya declared, pointing an accusing finger. "Heretic against the Great Continuance."

Mia stepped in front of the trolley, putting herself between Prya and the Shard. "This isn't about religion, Prya. The Shard is dangerous. It's warping our minds. Turning us into dangerous actors."

"Rubbish!" Prya countered. "It's divine revelation! Our ancestors have waiting for this. Sixteen thousand years of travel, all leading to this moment of transcendence."

One of the acolytes—a maintenance technician named Zavestki—spoke up in a high pitched voice that bordered on the shrill. "She means to cast away our salvation. Deny us passage to the next evolutionary state!"

Mia couldn't believe what she was hearing. She had already

told them the Shard was dangerous. What part of dangerous didn't they understand? "Don't you get it?" she said, injecting some of her own authority into her voice. "The Shard emits a bioelectric field operating at 40 to 100 Hz—the same frequency as human gamma brainwaves. It's creating neural synchrony that amplifies unconscious fixations. Maladaptive fixations. Hayes' violence, Dunbar's obsessive behavior, even Chen's experimental risks—they're all symptoms of exposure." There, she said it, verbalizing the idea that she had come to believe was the underling mechanism of the Shard's effect.

"Blasphemy!" the second acolyte shouted.

Prya raised her hand for silence, her eyes never leaving Mia. "I will broadcast your treachery to the entire crew. Let everyone know how Officer Hawthorn has attempted to destroy our future."

"This won't help anyone, Prya," Charlie intervened, his synthetic voice calm despite the escalating tension.

"The machine speaks," Prya said dismissively. "A tool questioning its own purpose."

Before Mia could respond, the bulkhead moved again. She tensed, anticipating reinforcements for Prya's side. The door slid open to reveal Kendricks, his security uniform crisp despite the hour, his facial expression bordering on triumph.

"You again," he scowled at Mia.

Prya turned proudly. "Security Chief, I caught them attempting to jettison the divine artifact."

Kendricks surveyed the scene, taking in the trolley with the tarpaulin-covered Shard, Charlie by the airlock controls, and Mia's defensive stance. Resting his eyes back on Prya, he raised his comms unit to his mouth.

"Kendricks to Captain Carson. Sir, we have a situation at the Master Airlock. Hawthorn and the simulant are making an unauthorized attempt to eject the artifact."

Mia's heart sank. With Carson involved, their plan was finished. The captain's authoritarian grip on the ship meant severe consequences—the Brig certain now, and reprogramming for Charlie, equally disastrous.

"This is for everyone's safety," Mia said, refusing to back down. "The Shard is creating cognitive distortions. It's affecting judgment, creating obsessions—"

"Salvation requires transformation," Prya interrupted, stepping closer to the trolley. "The old ways must die for the new to emerge."

Kendricks silenced them both with a raised hand, listening to his comms unit. After a moment, his expression changed subtly.

"Yes, sir. I understand." He lowered the device and looked directly at Mia. "Captain Carson is on his way. No one touches anything until he arrives."

An uncomfortable silence settled over the chamber. Mia exchanged a glance with Charlie, who stood perfectly still, his posture betraying nothing of his thoughts. Between them, the Shard hovered under the tarpaulin, a silent specter.

"The first humans who saw fire feared it," Prya said, breaking the silence. "Those who embraced it became gods among their kind."

"And those who worshipped it blindly burned," Mia countered.

Kendricks paced the perimeter of the chamber, keeping everyone in view. "Save the philosophy for the Dome debates. Captain's orders are clear."

Minutes stretched like hours. The stand-off settling into a waiting game. Finally, the bulkhead door opened once more, and Captain Carson entered, followed closely by Chen and, to Mia's surprise, Devory.

Mia looked at Devory, hoping to get at least some sympathy, but his expression, if anything, was one of disappointment.

Carson took in the scene with a sweeping glance, his weathered face unreadable. "Dr. Hawthorn," he said, voice even, "what is going on here?"

Mia realized she had to think quick, if she was going to have any chance of defending her actions. "I know this doesn't look good," she said, "but the Shard is not what people think it is. It's affecting our cognition and it's doing it without us realizing it, that's why—"

"And somehow you're not affected by it?" Carson cut in.

Mia felt like she was in a witness box being stared down by an unsympathetic jury. "I didn't say that," she said, hearing the meekness in her voice, despite herself. "What I meant was—

"The data doesn't lie," Dr. Chen interjected, stepping forward. "We've measured enhanced neural connectivity in all subjects exposed to the artifact. What Dr. Hawthorn calls pathological, I call breakthrough. The Shard is accelerating our cognitive evolution."

"And Hayes? Was murdering Laylee Watts part of this 'evolution'?" Mia challenged.

"An outlier case," Chen dismissed. "One negative outcome doesn't invalidate the potential benefits."

"And so you decided to take matters into your own hands?" Carson said, returning his attention to Mia, ignoring the exchange.

"Yes. To save the ship," Mia said simply.

Carson nodded, his face unreadable. "I'm afraid I have to agree with Kendricks. You've become a danger to the crew. You decided to act unilaterally without consulting your superiors. I have no choice but to hand you over to security."

"But—"

"Silence!" Carson ordered, his voice carrying the weight of sixteen thousand years of command hierarchy. "You have forfeited all your rights and privileges.

Mia wanted to argue further but recognized the futility of it.

"Security Chief," Carson, turning to Kendricks, "transport Hawthorn to the Brig."

Kendricks stepped forward and gripped Mia's upper arm. "I suggest you come quietly," he said, "unless you want to be restrained." He was referring to the pair of handcuffs hanging from his belt.

"It's okay," Mia said, while signaling her submissiveness.

Kendricks led her past Devory. Devory leaned in and whispered, "You tried to dispose of it without consulting me."

"You're affected, Dev," she said quietly. "We all are."

He shook his head, a mixture of hurt and confusion in his eyes.

Kendricks said to Devory, "Contact Rivel. Ensure that the Shard is moved to a safe location. I'll check in as soon as I've secured the prisoner."

"Understood, I'll take care of it."

Mia cast a glance back at Charlie as they crossed the threshold. She mouthed the words, "We failed," to him.

Charlie's amber eyes briefly dulled, then returned to their normal luminosity.

Kendricks clamped down on Mia's arm, forcing her through the airlock rampart. She didn't look back, but she felt the hostility of Prya's eyes burning into her back as she walked out.

To her surprise, Dr Chen came up beside her and said, "I'm afraid this is for your own good Mia. You haven't been yourself lately."

Mia looked at Chen and thought about telling her how wrong she was, but she could see it was no use. The doctor's pinpoint pupils and slight facial tic indicated advanced entrainment by the Shard's field. Kendricks too. He wanted to throw the Shard out just as much as she did, but here he was, following Carson's orders. Just what did Carson say to him? Or was this Kendricks' ploy all along, to ensure she was removed from the picture?

She felt the whole world had turned against her—everyone

except Tiber and Charlie. If what they were doing to her was any measure to go by, Tiber had better watch his back. If only she could get a message to him, but she felt it was hopeless. She felt tears welling up in her eyes, but she fought them off. If there was one thing she wasn't going to give to Kendricks, it was the satisfaction that he had defeated her. It was going to take a lot more than that to weaken her resolve.

Kendricks for his part seemed to be taking pleasure in parading his prowess. The normally officious demeanor he portrayed to the crew had been replaced by a smugness, revealing him to be nothing more than a brutish bully. She was no match for his physicality. If she was ever going to get out of this, she had to use her brain.

But what a choice! Her proximity to the Shard over the last hour meant that she was probably already compromised. It would only get worse. What shape that would come in, she could only guess. She just hoped that whatever it was, she kept her emotional core intact. If she could hang on to that, then all was not lost.

If only ...

— 21 —

THE BRIG

Kendricks forced Mia into the makeshift Brig with a callous shove. She stumbled forward, catching herself against the bare metal wall before turning to face him, her eyes blazing with defiance. The security chief stood in the doorway, his silhouette framed by the corridor lights behind him, his face half-shadowed and unreadable.

Mia looked around at what would be her prison for the foreseeable future. It was small, no more than several paces across in either direction. Not a proper detention cell at all, but an old equipment storage shed that had been hastily repurposed. It looked like it used to be a bone yard, where broken machinery was once cannibalized for spare parts. Evidence of its former purpose littered the floor—odd nuts and bolts, washers, errant wires, and control knobs scattered like archaeological remnants of the room's previous life.

In one corner sat two buckets; one for water; the other obviously meant to serve as a toilet. A narrow bunk with a threadbare blanket occupied one wall, its thin mattress garnered from some refuse heap. The lighting was harsh and unforgiving, casting everything in a sickly yellowish glow that made the metal walls appear jaundiced.

Dr. Chen, who had accompanied Mia this far, surveyed the conditions with growing discomfort. Her eyes darted from the buckets to the bunk to Mia's face, and something like shame flickered across her features. This was no way to treat a senior scientist, colleague, and until recently, a friend—no matter what rules she had broken. Chen's hand reflexively went to her pocket, withdrawing a protein bar.

"I'm sorry, Mia," she said, her voice barely above a whisper. "This is all I've got. I'll come back and check on you as soon as I can."

Mia's first instinct was to reject the offering. Pride and anger seethed within her, telling her to refuse this small kindness from someone who stood by while she was imprisoned. But Chen didn't wait for her decision. She hurriedly pressed the bar firmly into Mia's palm, squeezing her hand as if trying to communicate something beyond words.

Kendricks, impatient with this display of sympathy, shifted

his weight and cleared his throat. "Enough, Dr. Chen. Leave us alone."

Chen lingered for just a moment longer, her eyes meeting Mia's with a complicated mixture of regret and resignation, before she turned and exited without another word. The hatch hissed shut behind her, the sound of its pneumatic seal emphasizing Mia's isolation.

As soon as Chen was gone, Mia spun on Kendricks, fury erupting from her like a geyser that had been building up pressure. "You wanted it gone just as much as I did! You demanded its ejection at the council meeting. Don't pretend you're above this!"

Kendricks leaned against the wall, crossing his arms over his chest as if settling in for the entertainment. His flint-hard eyes glinted with mockery. "Carson needed to see the proof of justice done. You were the only one I could rely on to try something as stupid as this."

"Oh, you're a piece of work," Mia said, tapping her temple. "You're just as guilty as I am!"

Kendricks laughed, a cold, hollow sound that bounced off the metal walls. "That's why you're in here, Hawthorn, and I'll be out there." He gestured toward the door with a casual flick of his wrist, as if the difference in their positions was nothing more than an amusing coincidence rather than a calculated betrayal.

Mia stepped closer, every muscle in her body tense with righteous indignation. "You're a murderer and a tyrant. You won't get away with this, Kendricks!"

"I wouldn't get too excited if I were you," he replied, utterly unmoved by her accusation. He examined his fingernails with exaggerated interest. "I'd save your energy if I was you. You're only going to get one square meal a day while you're here. I can assure you. I'm in charge of the schedule."

Mia fumed, her breath coming in short, sharp bursts. Hunger was the last thing on her mind; her anger had sated that for now. Another part of her, however—analytical even in crisis—knew that she was powerless here. Attacking Kendricks verbally was exactly what he wanted. She could see it in his eyes. He was enjoying himself, feeding off her rage and frustration.

With a supreme effort born of years of scientific discipline, she quenched her emotions. She stepped back, deliberately unclenched her fists, and smoothed her features into a mask of indifference. Let him hang with nothing else to use against her. She would not give him the satisfaction.

Her sudden calm seemed to irritate Kendricks more than her anger had. His mouth twitched downward briefly before he recovered his smug demeanor. He took his time, languidly pushing himself away from the wall and dusting off his uniform as if their conversation had somehow soiled it.

"Get comfortable, Hawthorn," he said as he stepped back through the door. "I've got a lot of paperwork to fill out because of you. And I'm in no mood for it. I might just let it slip for a few days."

He went to the door and opened it, then stopped. Turning

around he said, "Oh yeah, one more thing Hawthorn."

Mia looked at Kendricks, unsure of his intentions.

Kendricks held his hand out. "Comms unit. Hand it over."

She should have known. Reluctantly, she unclipped it from her lapel and gave it to the security chief.

With that final act, Kendricks departed. The hatch sealed shut with a rusty clank, sealing Mia in alone with nothing but the hum of ventilation and her own tumultuous thoughts.

She stood motionless for several long moments, listening to Kendricks' retreating footsteps until they faded entirely. Only then did she allow her shoulders to slump, the weight of failure pressing down on her. She looked at the protein bar she still had in one hand. It was a good quality one. Dr Chen had privileges. She tucked it into her pocket, and laid down on the bunk.

For a while, she lay there with her eyes open, the yellow light shining annoyingly right over her. Eventually she turned to one side and curled up, shut her eyes and tried to shut her mind down.

But Tiber's words kept coming back to her. *He's sure to find a legal excuse to hang you.* What was the worst punishment for insubordination? Demotion. Rank reduced to a lowly maintenance worker. Or worse, a cleaner. But that assumed the legal process was run by sane individuals. And right now, that was far from being assured ...

With Mia safely locked away, Kendricks whistled to himself as he made his way through GenStar's corridors toward Devory's lab. The tune was an ancient one, something about kings and fallen empires that his grandmother had taught him as a child. It suited his mood—triumphant yet aware of the precariousness of power.

Crew members gave him wide berth as he passed, eyes averted, conversations pausing mid-sentence. Word of Mia's arrest had spread quickly, along with rumors of attempted sabotage. Kendricks did nothing to correct these misconceptions. Fear was a useful tool in maintaining order, especially now with the Shard's influence creeping through the ship like an invisible tide.

When he arrived at the GeoLab, he found Devory and Cory Dunbar laboring to rehang the Shard in its containment net. The artifact seemed heavier than its mass should allow, requiring both men's strength to maneuver it into position. Sweat beaded on their foreheads despite the lab's cool temperature, their movements jerky from exhaustion.

The Shard itself appeared unchanged—an irregular tetrahedron of impossible density, its surface a deep obsidian that somehow reflected nothing at all. Kendricks felt a momentary unease looking at it directly, as if his eyes were sliding off its edges in denial of its existence. He pushed the sensation aside. Weakness had no place in his position.

"Security Chief," Devory acknowledged without looking away from his task. His voice carried the brittle edge of someone stretched too thin by competing priorities. "Come to check on your runaway prize?"

"I'm securing this lab," Kendricks announced without preamble, ignoring the scientist's attempt at sarcasm. "Extra monitoring, restricted access. From now on, all personnel entering and exiting this lab will be required to log in to Security. Is that understood?"

The Shard settled into its net with a final push from Dunbar, who then stepped back, wiping his palms on his jumpsuit. Like Devory, his eyes remained fixed on the artifact rather than on Kendricks. Both men nodded in unison, the synchronicity of their movement almost eerie in its precision.

Kendricks waited till both men left the lab, Devory locking it behind them.

As they passed, Kendricks noted the geometric patterns Dunbar had scratched into his own forearm—perfect trapezoids repeated in rows, some fresh enough that tiny beads of blood still welled from the scratches.

Something about Dunbar's behavior disturbed him, the Hayes' incident still fresh in his mind. He made a mental note to pass this information onto Lora Vasquez. Lately, she never seemed to be around when he needed her. He decided that she was probably tied up with her usual beat in the Pleasure Quarters. There hadn't been any more incidents, so he assumed she had it under control. Nevertheless, he would pay her a call as soon as he had this other business wrapped up.

His next stop was the Archives.

Meanwhile, in the Archives, Charlie stood before Tiber, his

titanium-alloy frame somewhat agitated compared to his normally efficient and calibrated movements. His amber eyes dimmed and brightened in irregular patterns, the simulant equivalent of human distress.

"I messed up," he lamented, his voice modulating more randomly than usual.

Tiber nodded grimly from his seat among the data terminals and memory cores that formed the heart of the Archives. The old historian's face was mapped with deep lines, each one a record of some crisis weathered, some difficult truth preserved. The current situation was etching new ones around his eyes.

"There's nothing that can be done about it now," he replied, fingers idly tapping a rhythm on the armrest of his chair. "All we can do is wait for the consequences."

As if summoned by these words, the entrance door to the Archive hissed open with more force than usual, announcing Kendricks' arrival before he physically appeared. The security chief strode in with military precision, his boots striking the deck plates with sharp reports that echoed through the cavernous space.

He walked briskly to where Tiber was sitting at his desk. Charlie, for want of a better expression, was cowering on the other side, his mechanical body hunched slightly as if trying to make himself less visible.

Kendricks pointed accusingly at Charlie, raw anger evident in the rigid line of his jaw. "You let your dog loose," he snapped at Tiber. "If I find out that you helped Hawthorn in any way, I'll see to it that all your privileges are revoked. You can log that in your notes."

Tiber's weathered face tightened, but his historian's calm held. Centuries of maintaining neutrality in the face of GenStar's political shifts had taught him the value of a measured response. "I'm an archivist, Kendricks, not a conspirator. Charlie acts on his own logic."

"Logic?" Kendricks sneered, the word dripping with contempt. "He's a tool, and you're his keeper. Keep him leashed, or I'll have him scrapped." He turned to Charlie, jabbing a finger toward the simulant's chest plate. "Anna Teoh's coming to reprogram you. Leave this room and you're done."

The threat hung in the air like a physical presence. Charlie remained silent, his processing systems calculating thousands of potential responses and discarding each one as inadequate.

Kendricks glared at them both for a moment longer before storming out, the hatch sealing with a clang that reverberated through the Archives like the final note of a funerary dirge.

Tiber exhaled deeply, rubbing his temples with gnarled fingers. "I'm afraid we haven't got much time," he said, his voice lowered despite the security chief's departure. "We need to figure a way to get Mia out of the Brig."

Charlie straightened, his mechanical posture returning to its optimal configuration as his systems focused on problem-solving rather than self-recrimination. "I've been running simulations," he said. "The only option that I can see having any chance of success is to enlist the help of Devory."

"But didn't you say he had turned?" Tiber asked, confusion evident in his furrowed brow.

"I did, but based on the scans I did of Mia and Dr. Chen, there's a remote possibility that his emotional circuits are still intact."

"Explain," Tiber said, not exactly following the simulant's reasoning.

Charlie moved closer to the terminal where Tiber sat, his movements more fluid now that he had a purpose to fulfill. "The Shard's effects appear to only go cortex deep—at least that is what the data showed at the time I observed Mia and Dr. Chen's tests. Mia remarked that their limbic systems, the areas responsible for emotional memories, had not yet been affected."

He paused, head tilting slightly as he processed additional variables. "Perhaps it had something to do with distance and time since exposure. Time therefore is of the essence. If we are to have any chance of helping Mia escape, we need to approach Devory right away."

"Okay," Tiber said, nodding slowly as he considered the implications. "But if we do that, we need to do it the right way. We need a plan."

"Yes. What do you suggest?"

Tiber rose from his chair, moving with the careful deliberation of age toward a holographic display of GenStar's crew relationships and hierarchies—a sociological map he had maintained for decades. He highlighted the connection between Mia and Devory, which pulsed with data points indicating emotional entanglement.

"We know they have affection for each other," he said, studying the visualization. "We need to trigger Devory's memory of that. But in order for us to do that so that he doesn't just dismiss us, we need something concrete to anchor him. Something that will engage his limbic system over his cortically dominant thoughts."

"You mean feelings over logic," Charlie translated, his processing capacity making the leap instantly.

"Yes."

Charlie's amber eyes brightened as he accessed his memory banks. "Mia told me he is obsessed with finding other pieces of the Shard," he said. "I know for a fact that Dr. Chen has one such fragment in her laboratory. If we can somehow lay our hands on that, we could bring it to Devory and say that we found it lodged in the asteroid regolith somewhere."

The simulant's voice modulation shifted slightly, indicating increased confidence in his proposal. "When he asks us where, we can lead him to the Brig, and from there let Mia try and persuade him to release her."

Tiber stroked his chin, considering the plan's merits and risks. After a moment, he nodded decisively. "I have just the idea," he said, a rare sparkle of mischief lighting his ancient eyes. "I will personally take care of Dr. Chen. Wait until I have her occupied in her office. Then get the fragment from her lab."

Charlie's servos whirred softly as he processed this unexpected development. "You? But how will you distract her?"

A smile spread across Tiber's face, transforming him momentarily from austere archivist to cunning conspirator. "I've been recording the medical history of this ship for over six decades. Chen may have forgotten, but I haven't. She and I have unfinished business from the 3991 Blowout incident—records she'd rather not see made public. It's time to suggest a review of that particular archive."

"Blackmail?" Charlie asked, his programming struggling to reconcile this approach with ethical parameters.

"Historical leverage," Tiber corrected mildly. "For the greater good."

Charlie considered this, his ethical subroutines running comparative analyses against potential outcomes. After 3.7 seconds, he reached a conclusion. "The needs of the many outweigh the needs of the few," he quoted, an ancient reference pulled from his cultural database.

"Precisely," Tiber agreed, already moving toward the Archive exit with more energy than he had shown in years. "Be ready to move when I signal you. We have one chance at this."

As the archivist departed, Charlie turned his attention to preparing for their mission. His systems calculated a 37.2% probability of success—low, but not impossible. For Mia's sake, they would have to beat the odds.

Back in the makeshift Brig, Mia paced the confined space, five steps in each direction before being forced to turn around. The protein bar Chen had given her was gone, devoured in small, measured bites despite her initial resistance. Her stomach already growling for more.

The lighting remained constant, offering no indication of Gen-Star's artificial day-night cycle. Without her comms unit or access to ship systems, time blurred into an indistinct stream of moments, measured only by her increasing thirst and the gradually filling waste bucket in the corner.

She had examined every inch of her prison, cataloging potential weaknesses or tools. The scattered hardware on the floor yielded nothing useful—odd pieces too small or broken to serve any practical purpose. The ventilation grate was too small for anything but her arm to fit through. The door mechanism was external, with no access panel on her side.

Frustrated but not defeated, Mia returned to the bunk and sat cross-legged, closing her eyes to conserve energy and focus her thoughts. If physical escape was impossible for now, she would have to prepare mentally for whatever opportunity might arise.

Her father's final mission had begun much like this—a hopeless situation, deteriorating conditions, impossible odds. Yet he had found a way forward, sacrificing himself to save others. *Firmate Catena Maiorum*. The chain of ancestors stretching back to Earth itself, each link forged in crisis and sacrifice.

"I won't break the chain," she whispered to herself, fingers tracing the outline of the medal through her jumpsuit. "I'll find a way."

– 22 –
THE MOB

A large crowd had gathered around the Common Dome, waiting to listen to Prya Chandra's latest announcement. Rumors had been circulating, and people had begun converging on the Dome with torchlights after emerging from the dark tunnels that radiated off it like spokes of a wheel. The Dome reflected the torchlight in a crisscross pattern reminiscent of a kaleidoscope, giving the space a semi-surreal atmosphere.

Prya stood on a platform improvised from agri-crates stacked against the Dome railing, her slight frame radiating unshakable conviction, her face glowing with an ethereal other-worldliness from the artificial illumination. In her hands, she cradled her Shard fragment, now encased in a ceremonial holder of polished titanium, its lattice etched with geometric runes—interlocking trapezoids spiraling inward toward infinity. The fragment swallowed the light, its surface a void that defied logic. Yet those standing meters away felt their eyes drawn to it, pupils dilating as they stared.

Devory stood at the edge of the crowd, his face drawn from the exertions of returning the Shard to his Lab. He watched Prya with a mixture of scientific fascination and growing unease—part of him seeing her as competition for the Shard's secrets, another part recognizing, somewhere in his fragmenting consciousness, that what was happening before him represented exactly what Mia had warned him about.

Not the same could be said for Cory, who stood next to him and agitated with what seemed like an abundance of energy directed at worshipful devotion. He made the extra effort of rolling up his sleeves so that he could proudly show off the geometric shapes he had carved into his flesh.

Prya raised the holder, and the crowd fell silent, their eyes locked on the artifact's uncanny darkness. "The Shard has opened my mind to visions!" she declared, her voice clear and fervent, cutting through the crowd's noise. Where she had once merely been a conservative spiritual advisor, she now spoke with the authority of a self-proclaimed prophet. "I see the future, and it is not on Eridani-B. The mission commanders lie! The planet is a death trap, far more hostile than they admit. Its atmosphere teems with microscopic silicates that will shred our lungs! Its water harbors bacteria that will dissolve our bones! Colonization will end in ruin, our bodies broken on its unforgiving crust!"

Gasps rippled through the crowd. A technician clutched a trapezoid-etched talisman hanging from a cord around her neck, echoing Cory Dunbar's geometric fixation. Behind her, a miner traced the same shape repeatedly on his palm with a calloused finger. Prya's words stirred their minds, urging surrender to the Shard's psychological pull, its influence subtle yet infinitely powerful in the way it amplified unresolved unconscious motivations.

"She's seen it!" someone whispered. "The Shard shows her the truth!"

"But Carson said the atmospheric readings were stable," a nearby woman in a faded blue jumpsuit countered, only to be silenced by glares from those around her.

Prya's eyes blazed with certainty as she paced the platform, her movements deliberate, theatrical. "The future is here, on this ship, with *this* Shard!" She thrust it in the air triumphantly. "It calls us to new work—we shall hollow out the core, carve space for gardens, homes, life!" She gestured expansively, encompassing the entire Dome. "Imagine this space, ten times larger! Where food grows in abundance, where children play without rationed water and oxygen, where we are no longer the servants of a dying mission but masters of our own destiny! A renaissance of growth and prosperity awaits, where scarcity ends, and our children thrive under the Shard's light!"

Her voice soared, painting a tangible dream, a ship reborn, its decay reversed. The crowd leaned forward, some weeping, others nodding, their doubts dissolving in the promise of a concrete salvation. Parents pulled children closer, envisioning futures where vitamin D deficiency and protein shortages were distant memories.

"But now I shall tell you something you all need to know," Prya cried, her tone subduing to righteous fury. "There are heretics! They threaten us as we speak!" She slammed her fist against the platform railing, the strength of the physical impact belying her slight form. "Mia Hawthorn, a scientist turned traitor, conspired with Tiber's simulant, Charlie, to eject the Shard—our guide! They sought to doom us to Eridani-B's grave!"

She pointed dramatically toward the security section where the Brig was located. "Even now, she plots against us. She would rather see your children gasping for breath on a hostile world than accept this gift the universe has bestowed upon us!" She shook the fragment again.

Gasps turned into angry cries. A miner, gaunt from ration cuts, shouted, "Heretic!" The chant spread, venomous and unified, echoing off the Dome's curved surface, amplified by the acoustic properties of the space.

"She deserves worse than the Brig!" shouted a woman whose sleeve patches identified her as part of Agriculture.

Prya's lips curled into a snarl, her grip steady on the cold titanium holder. "Hawthorn festers in the Brig, but her allies scheme. Yes! Tiber, the archivist, harbors the simulant who betrayed us."

She leaned forward, lowering her voice so the crowd strained to hear. "And what do you think they discuss in those Archives? What plots do they hatch while we struggle to survive?"

The crowd murmured angrily, their collective fear focusing on this new target.

"What shall we do?" Prya asked, her voice rising again.

Zavetski, the shrill acolyte from the Master Airlock standoff, pushed to the front of the crowd. Her jumpsuit was now adorned with geometric symbols cut from thick aluminum foil and sewn to the fabric. She screamed, "Destroy the Archives!"

The cry ignited the crowd, their voices merging into a deafening roar: "Destroy the Archives! Destroy the Archives!" Fists pumped, eyes gleamed with the Shard's enigmatic geometry, their thoughts warped by its invisible influence.

Rivel emerged from an entrance tunnel and stood at the back of the room, curious to see what was going on. The moment she saw Prya, her face contorted with hate. She scanned the feverish crowd, searching for Carson and Kendricks among the sea of entranced faces, but neither was present. Her hand instinctively moved to the utility knife clipped to her belt, a move she would have never dreamed of making until this moment.

The crowd's fervor had transformed the usually ordered Dome into something unrecognizable—these weren't her crewmates anymore but strangers caught in a collective madness. She spotted Devory at the edge of the gathering, his face a mask of conflicted emotions, and beside him Cory, displaying his self-mutilation like badges of honor. Others she had shared meals with just days ago now chanted for destruction with vacant, synchronized expressions.

She wanted to act, to challenge Prya's dangerous rhetoric, but logic prevailed. The crowd would tear her to shreds before she could even reach the platform. Reluctantly, she backed away, melting into the shadows of the tunnel she had come from, mind already racing with plans to warn Tiber and Charlie.

As soon as she was out of sight, Rivel broke into a sprint through the maintenance tunnels, taking the fastest route to the Archives. Her boots echoed against the metal flooring as she pulled out her comms unit, frantically trying to reach Tiber.

'Tiber, come in! Charlie! Anyone!' she called into the device, only to be met with static. The comms systems must have been affected—whether by technical failure or deliberate sabotage, she couldn't tell. Communications blackouts weren't uncommon on GenStar, but the timing couldn't be worse.

She pushed herself harder, lungs burning as she took a service ladder down three levels, skipping rungs. If she couldn't warn them remotely, she'd have to get there first. The Archives had minimal security—it had never needed it before. Against a mob like that, its door wouldn't hold for long.

Rounding a corner, Rivel collided with a maintenance worker who stumbled back in surprise.

'Hey, what's the—'

'No time!' Rivel cut him off. 'There's a mob heading for the

Archives. Get Security on comms if you can!'

She didn't wait for a response, already moving again, calculating the routes in her head. The mob would take the main corridors—she might still beat them if she continued through the service tunnels. But even as she ran, she heard the distant thunder of footsteps and angry voices echoing through the ship's skeleton, moving with terrible purpose.

It seemed Prya had caught a glimpse of her, and shouted, "That's right! Leave you traitor!" The crowd booed in response. "See! We shall cleanse the ship," she cried, "starting with the Archives. Destroy the old narratives. Down with the dying mission! A new era begins today!"

The mob took this as their signal and surged through the tunnels shouting, "Down with the Archives!" A flood gate had been opened. The narrow passageways, designed for efficiency rather than crowd control, funneled them into a dense, unstoppable mass. Overhead lighting flickered as they passed, as if the ship itself trembled at their fury. None noticed or cared that Rivel had already disappeared, racing against them through the ship's veins.

Prya led, holding the Shard fragment aloft, its light-swallowing surface a stark contrast to the vial's glint under flickering lights. Acolytes—Zavetski and Brenner, the latter a wiry engineer with trapezoid-scarred arms—flanked her, their devotion fierce. Brenner carried a heavy magnetic drill, its power cell warning light blinking orange—nearly depleted, but enough for destruction.

"Make way!" Zavetski shouted at a group of off-duty crew

members who stared in confusion at the approaching mob. "The Priestess of the Stone is here!"

Hesitant crew members were swept along or shoved aside, their skepticism crushed by the chant. A young technician who questioned where they were going was knocked against a wall, his protest silenced. The Shard's psychological grip amplified their emotions—hope, rage, unity—mirroring the mass hysteria of the Long Night collapse centuries ago. History repeating itself in the echoing corridors of humanity's great experiment.

The mob's footsteps thundered, drowning all sensible thought. A maintenance worker swept up in the current muttered, "Worse than the Dome riot." His colleague squeezed along with him voiced aloud, "This is a bloody revolution!"

The mob turned a corner and descended a ramp that spiraled into the lower levels of GenStar, where the Archives were housed, protected by distance and architecture from the ship's more vulnerable sections. On the walls, centuries-old markings from the original Arkonauts could still be seen, their faded symbols a testament to how far humanity had traveled, both in distance and in spiritual evolution.

Rivel arrived at a service entrance to the Archives, breathless and desperate. She punched in her override code, but the panel flashed red. Someone had already locked down the peripheral access points.

"Tiber!" she shouted, pounding on the door. "Charlie!" Her voice echoed in the empty maintenance corridor. No response.

She pressed her ear to the cold metal, straining to hear movement inside, but detected nothing. The Archives were vast, perhaps they were in a distant section and couldn't hear her? Or perhaps they weren't there at all.

A distant crash followed by cheering redirected her thoughts to her own personal safety. The mob had arrived at the main entrance. She was too late.

With no weapons beyond her utility knife and no way to stop dozens of frenzied crew members, she made the tactical decision to retreat. She needed to find Carson, Kendricks—anyone with authority who might help. With a last regretful look at the sealed door, she slipped back into the maintenance tunnel, the sound of pounding feet drumming in her ears.

Moments later the frenzied mob reached the entrance to the Archives. They were forced to stop in front of its large entrance doorway. Etched above its lintel were the words "VERITAS OMNIA VINCIT," and it gave them a slight pause, as if entering this hallowed place might incur an unspoken wrath. The mob quietened down, waiting for Prya's guidance.

Prya lowered the fragment, its inert void seeming to drink the corridor's light. For a moment, doubt flickered across her face. Was this the true path? Was she in control? But then a spasm crossed her face, a momentary glimpse of madness, and her thoughts crystallized like frost on glass.

"For the Continuance!" she shouted, the irony lost on her that she invoked the very historical mission she sought to destroy.

The mob surged forward, wielding tools and fists. Brenner stepped in front with his magnetic drill, its bit spinning against the steel plate door, sending sparks flying. Others joined with metal bars and makeshift battering rams constructed from sections of pipe. The hatch groaned, its locking mechanism—designed to protect against accidental damage, not concentrated assault—failing under the onslaught.

The door buckled, then crumpled open to reveal the Archives' cavernous interior: shelves of data tablets, memory cores, GenStar's historical heart. The smell of musty paper and the metallic scent of ancient electronics wafted out, the odor of history itself.

Tiber and Charlie were nowhere to be seen. But this didn't stop the mob from carrying out their violent intentions. If anything, it gave them license to fully unleash their madness, without the uncomfortable presence of witnesses who might question their righteousness.

The Archives stood silent, defenseless. Row upon row of data storage units—some ancient technology from Earth itself, others fabricated during the voyage—gleamed dully in the emergency lighting that had activated when the door was breached.

"Cleanse it all!" Prya commanded, stepping over the threshold. "Let nothing of the old narratives remain!"

Zavetski, her eyes blazing with fervor, smashed a terminal with a wrench, sparks bursting as circuits fried. The display flickered, showing fragments of text—"Earth transmission 9924 ... final contact ..."—before going dark forever. Brenner toppled a shelf, data tablets shattering like brittle bones, their fragments dusting the air with the remnants of humanity's journey.

A miner clawed at a memory core, cables snapping like tendons, while others crushed consoles, their screams of "Heretic!" melding with the screech of metal and the crack of plastics. Someone found an ancient paper book—a rarity, perhaps the last of its kind—and tore its pages, sending them fluttering through the air like dying butterflies.

The air choked with the acrid stench of burnt polymers and the dust of centuries, the Archives' sanctity—16,000 years of records, including the Long Night's collapse—reduced to rubble. History wiped in minutes that had taken countless man-hours to document, memoirs and logs and scientific observations turned to ash and broken silicon.

In the center of this destruction, a small holographic projector, somehow still functioning, displayed a rotating image of Earth— blue and green and white, pristine as it never truly was. No one noticed it, or if they did, they ignored this ghost of humanity's cradle.

Prya stood amid the chaos with arms raised, holding the fragment up, a Priestess of Chaos. The emergency lights caught the edges of the holder, sending prisms of color across her face, transforming her features into something otherworldly and terrible.

"The heretics' lies are broken!" she declared, looking around at the damage through the dust-filled air. Bits of data storage media crunched beneath her boots as she walked to what had been Tiber's main console. Placing the Shard fragment atop it, she gazed out at her followers, their faces streaked with sweat and grime, their eyes reflecting the madness she had unleashed.

"A new dawn awaits us!" she proclaimed. "It is here!" And the mob shouted in unison after her: "A New Dawn is here!"

— 23 —
RESCUE

The Archives lay in jumbled ruins. It looked as though vertebrae had been snapped from a spine and scattered across the room. Emergency lights flickered, casting jagged shadows over the wreckage of GenStar's memory. The dust hung in the air. It settled on Charlie's titanium outer shell as he stood motionless, a shadow among the devastation. In his hand, the stolen Shard fragment, sealed in its vial, lay dormant—its obsidian surface swallowing the light, yet heavier than anything he was accustomed to, forcing him to recalibrate his sensory feedback system. His amber eyes tracked the last remnants of Prya's mob as they picked their way through the devastation, the first flickers of shame crossing their faces as they realized what they had done: sixteen millennia of history—every memoir, scientific log, and cultural record—reduced to rubble. They didn't glance at Charlie. To them, he was just another broken object.

Footsteps echoed at the entrance. Devory emerged, his eyes taking in the destruction with disbelief. Somewhere in the fractured mirror of his own mind, he recognized that forces had been unleashed that he was only too dimly aware of.

Behind him, Anna Teoh, navigator's assistant, clutched her glowing data tablet, its light cutting across her face. Her sharp gaze swept the ruins, lips tightening in silent outrage at the loss of GenStar's legacy. Charlie tilted his head, his amber eyes widening with a welcoming recognition.

"Devory," he said, voice cutting through the haze, "I have a fragment for you. I can show you where others can be found."

Devory's brow furrowed. He stepped closer, eyes locking on the fragment's light-swallowing surface. "That's the one I gave Mia," he said. "Why are you lying, Charlie?" He humphed deliberately to show his displeasure.

Charlie maintained his outward poise. His neural net calculated that lying would trigger an emotional response. How much, he wasn't sure. But it seemed to be enough. He had opened a channel. "Because I need you to save Mia," he said sternly.

Devory stared at Charlie for a moment, as if he didn't register the simulant's words. But then something flickered, a memory, and his eyes changed. "So that's what you're asking," he said, nodding slowly.

"Yes," Charlie said, indicating the destruction around them. "It has come to this."

Anna had listened to this conversation with patience, but she began to get restless. "Charlie you realize I've been sent here to reprogram you?"

Charlie's amber eyes flashed briefly with fear. "Yes," he said. "Kendricks told me you would come."

"He told me you and Mia tried to eject the Shard. Is that true?"

"Yes."

Anna shook her head in dismay. "Charlie, what were you thinking?"

"I believe Mia is right. I still do. So if you want to reprogram me, go ahead." He slowly turned around and exposed the data socket at the back of his head.

"Charlie!" Anna said, reaching out and touching his arm. "You don't have to do that." She was genuinely sympathetic.

Charlie turned around to face her again. "Are you saying you're changing your mind?"

"We've already lost enough history today Charlie. Wiping you memories would only make it worse, not better. And I can't be part of that."

"I'm glad to hear it," Charlie said. His eyes flashed a sunflower yellow, his version of happiness.

"I'm glad to hear it too Charlie," Devory said. "Because without you, I might have given up on Mia. But you reminded me of what I lost."

"And what is that?"

"Love."

Anna looked at Devory with fresh eyes. The rough and ready miner baring his soul.

"We need to rescue her," Devory said, looking at both Charlie and Anna.

"Rescue her?" Anna said.

"Yes," Devory said. "Kendricks took her to the Brig."

Anna pocketed her tablet, her expression resolute. "In that case, I'm coming with you," she said. "I know the mainframe systems—might be able to override the lock." Her navigator's training granted her access to GenStar's digital underbelly, a skill now wielded as defiance against Prya's cult.

The trio moved swiftly through GenStar's polygrout-lined corridors, the ship's artificial gravity gradually getting weaker as they descended deeper into the bowels of the asteroid. Shadows danced under flickering lights, mirroring the crew's fractured unity. Charlie spoke as they navigated, his voice cutting through the tension. "Mia still loves you, Devory. She believes your core self is still intact, despite the Shard's pull." Devory's jaw tightened, the words igniting a tangle of hope and guilt. His miner's logic, once unyielding, wrestled with the Shard's lingering distortions, amplifying his unresolved longing for Mia.

They reached the Brig. Mia's cell, a cramped room with a flickering overhead light and the faint hum of a failing ventilation unit, loomed ahead, its keypad glowing red, unyielding. Anna crouched, tablet linked to the mainframe, her fingers a blur over the interface. Charlie fed her override codes, his neural net parsing stolen data from Chen's lab with precision. The air thickened with anticipation, the distant chug of air recyclers a low heartbeat in the silence.

The lock clicked, a sharp sound breaking the silence. The door slid open to reveal Mia, pacing feverishly in her cell. Her hands were trembling, her face drawn with a mixture of desperation and determination. She was speaking to the air, to voices that only she could hear.

"Seneca ... you overcame the Long Night," Mia whispered, her voice cracking. "You rebuilt the trust that had been shattered. You survived when everything around you collapsed. You held the pieces together when all seemed lost." She stopped pacing, her wide eyes staring into the cold walls of her cell, as though she could see Seneca's presence there. "Tell me, how did you do it? How did you survive the darkness? How do I survive this? How do we overcome the Shard? How do we bring the ship back from the edge?" Her hands clenched into fists as she stood still, waiting for an answer that would not come. The voices in her mind, whispers from long-dead ancestors, seemed to grow louder in her ears, but there was no guidance. She needed Seneca's strength, the wisdom of the one who had witnessed the horrors of the Long Night and emerged to rebuild what was torn apart. "Teach me," Mia's voice softened. "Teach me how to bring us back from this madness."

Devory's heart lurched, a visceral ache cutting through his own distorted perceptions. He crossed the cell in two strides, and wrapped his arms around her, pulling her into a fierce embrace. Mia flinched, caught in Seneca's vision, her body rigid as if tethered to the Long Night's horrors. "Seneca, no—show me the way!" she gasped, but Devory held her tighter, his warmth penetrating their clothes. The steady rhythm of his breath, the earthy scent of his skin—sweat and regolith dust—grounded her. He buried his face in her hair, inhaling deeply, the soft fragrance piercing the fog. His senses, clouded by its influence, began to clear, each breath tethering him to reality. "Mia," he whispered, voice hoarse. "You're bringing me back. Let me bring you back."

She looked up, confusion and desperation warring in her eyes, Seneca's image fading like static. "Devory … the Shard—it's the Long Night again," she said, her voice breaking. "Seneca warned me … it's unraveling us, the ship. We have to stop Prya."

Devory wanted to answer, but a shrill alarm shattered the moment, piercing the cell's confines. Charlie's optics flared red. "The door triggered a Security alert. Kendricks will come hunting."

Anna's tablet pinged, its screen illuminating a labyrinthine map of GenStar's tunnels. "Service tunnel, now!" she cried. The determination in her voice seemed to shock them into the present.

She led them to a narrow hatch concealed behind a maintenance panel. As navigator's assistant, Anna had memorized every conduit, running them for thrills in quieter times. Now, her knowledge was their salvation. They crawled through the claustrophobic passage, polygrout walls scraping their shoulders, Anna's tablet guiding them past junctions. Her calm voice masked the tension. "This track avoids the main corridors. Kendricks won't expect it."

The four fugitives entered the Agricultural service tunnel, its air humid with the scent of hydroponic runoff and algae vats. Anna hesitated, her tablet revealing a riskier path. "Sub-service conduit," she whispered, diverting them into a narrower passage, barely wide enough for Devory's broad frame. "This leads to the gymnasium." They emerged near the swimming pool, its chlorinated tang sharp in the stale air, the indoor tennis court a ghostly expanse under halogen lights. Another tunnel plunged them deeper, into an old mining sector—a labyrinth of abandoned shafts where the Recyclers, GenStar's "alternative" Arkonauts, had carved a defiant, vibrant existence.

The Recyclers' enclave was a chaotic symphony of survival and ingenuity, a sprawling bazaar etched into the bones of GenStar's abandoned mines. Stalactite-like conduits dripped condensation, their moisture harvested by makeshift gutters feeding hydroponic patches of algae and fungi. Lanterns—cobbled from scavenged reactor filaments—cast a patchwork glow, illuminating stalls piled with upcycled treasures: circuit boards fused into abstract sculptures, polygrout shards polished into blades, salvaged pipes reborn as structural beams or musical chimes that hummed in the recycled air currents. Bartering voices clashed with the clank of tools, a dialect of trade layered with Arkonaut slang preserved from the ship's founding. Recyclers, their jumpsuits patched with reclaimed fabrics, bore trapezoid-scarred tattoos—echoes of the Shard's influence, adopted as badges of rebellion. A communal forge roared at the enclave's heart, its heat tempering polymer tools, while a nursery echoed with children's laughter, their toys carved from reactor shielding. A storyteller, her voice rich with cadence, wove Arkonaut myths of the Continuance to wide-eyed listeners, her words a haunting echo of Mia's visions. The air carried the tang of solder, the musk of cultivated fungi, and the faint ozone of overworked power cells, a testament to their self-sustaining ecosystem.

Eyes turned to Charlie, his synthetic frame gleaming like a prize for "reclamation." A wiry Recycler, her tattoos pulsing under bioluminescent ink, stepped forward, her gaze predatory. "That simulant's worth a month's rations," she said, voice low. Anna raised a hand, her navigator's authority cutting through the murmurs. "He's not for sale," she declared, her tone unyielding, eyes locking with the Recycler's until the woman backed off, muttering.

Vesta Torricelli emerged from the crowd, a weathered figure with silver-streaked hair and eyes sharp as a plasma cutter. Scars crisscrossed her hands, each a story of salvage and survival. Her patchwork jumpsuit jangled with repurposed artefacts, and hanging around her neck was an Arkonaut medallion reforged into a pendant, its faded runes glinting under the lanterns. "You're running from trouble," she said, her voice low. "Come." She led them through the enclave's maze, past the forge's molten glow, past a hydroponic vat where algae bloomed in vibrant greens, past the storyteller's circle where children clutched Continuance talismans. The fugitives' footsteps echoed on the uneven floor, a mosaic of welded scrap and polished stone, each step a reminder of the Recyclers' defiance of GenStar's decay.

Vesta watched them with sharp eyes. She seemed calm, but taut as a wire. Sure enough, there was the sound of commotion outside.

She made a low whistle with her lips and called over one of her brethren. "Hide them behind the cooling loops," she ordered.

The recycler led the four of them away to hide.

Vesta stood still with her hands on her hips.

It was Lora Vasquez, Kendricks' deputy. Kendricks himself would never come this far. He left the "dirty work" to his subordinate; it was beneath him. Vasquez approached Vesta. They weren't strangers. Vasquez often relied on Vesta to keep her informed of life in the Recycling tunnels. Many of the children never went to school. Vasquez would have preferred that to change, but she was fighting generations of tradition, as she was now facing Vesta.

"The tunnel heat sensors indicated a spike in this sector," she said. "Is there something you want to tell me?"

"Nothing to tell," Vesta said calmly.

"You think Kendricks won't notice you're harboring fugitives?" Vasquez's voice cut through the dim, rhythmic hum of the enclave as she neared. Her presence loomed, as solid as her reputation for tracking down any trace of dissent.

Vesta met her with the calm demeanor of someone who had lived through worse. "Still happy to be his lapdog I see," she said, picking a morsel of food from her teeth.

Vasquez stepped closer, eyes narrowing slightly as she searched Vesta's face for any sign of deceit. The tension between them was palpable, thick like the dust that clung to the underground walls. The fact that Vasquez was here, so far from the upper decks, spoke volumes about her commitment—or perhaps her desperation.

"You're playing a dangerous game, Vesta." Vasquez's words were low, calculated. She moved even closer, her hand drifting instinctively toward the weapon at her side.

Vesta didn't flinch. "We're all playing dangerous games in this place, Vasquez," she said, her tone still casual. She gestured toward the enclave around them, its labyrinthine design, the relics of old lives. "And yet here you are, looking for trouble."

"Hand them over and there won't be any trouble."

Vesta threw her head back with a casual laugh. "You never learn, do you? Read my lips. Go home."

Vasquez looked around. A group of recyclers had formed a ring around her. They held an assortment of pipes and jagged instruments, and looked hungry to use them.

Vasquez held her ground. "You think you can intimidate me?" She unsheathed her weapon.

"No," Vesta said, quietly. "But I'd think twice if I was you." And she reached across and picked up a crossbow.

Vasquez weighed up the odds. "Ok," she said, and she slowly backed away. "There's no need for that."

"Good dog," Vesta said.

Vasquez continued backing away. Some of the recyclers spat on her as she went.

After she was gone, Vesta took the four fugitives to her quarters, a hollowed-out mining chamber, was a microcosm of Recycler ingenuity. The walls, reinforced with salvaged bulkhead panels, bore etchings of Arkonaut star charts, their constellations a faded prayer for a lost Earth. A cracked holo-frame cycled images of oceans, forests, and long-extinct cities, its projector humming on a scavenged power cell, casting fleeting shadows across the chamber. A dented Arkonaut medallion hung above a workbench cluttered with tools: a plasma welder, a neural interface salvaged from a defunct simulant, a stack of polymer crates repurposed as shelves holding fungal protein bars, vials of recycled water, and a rare paper journal—its pages yellowed, possibly smuggled from the Archives before its fall, its cover etched with Continuance symbols. A curtain of woven circuit fibers partitioned a sleeping alcove, its bedding patched from jumpsuit scraps, exuding a faint scent of recycled polymers. A small hydroponic rig in the corner grew luminescent fungi, their glow supplementing the lanterns, bathing the space in an eerie, organic light. The air was warm, scented with the earthy musk of fungi, the metallic tang of Vesta's forge, and a hint of ozone from her power cell, a stark contrast to the Brig's sterile chill.

The three human fugitives collapsed onto a bench of fused polygrout, their breaths ragged from the escape. Mia, still trembling from her visions of Seneca Ramujan, leaned into Devory, her fingers tracing the seams of his jumpsuit, grounding herself in his presence. Her eyes darted to the paper journal on Vesta's shelf, its Continuance symbols stirring a pang of connection to Seneca's warnings. Devory ate mechanically, each bite anchoring his clarity, his free hand never leaving Mia's. Anna set her tablet on the workbench, its map dimming as she rubbed her eyes, exhaustion seeping through her navigator's resolve. She ran a hand through her bob-style hair, the warmth of Vesta's quarters easing the tension in her shoulders. Charlie stood by the door, his optics scanning the chamber, while his neural net catalogued Vesta's artifacts for potential utility—particularly the telephone system, which could eavesdrop on the GenStar's comms loop if needed. The warmth of the space, the hum of the holo-frame, and the distant clatter of Recycler work enveloped them, a fleeting sanctuary from their flight.

– 24 –

FAILSAFE REDUX

It was some time before the adrenaline had sufficiently flushed from their systems that Anna stood up and declared she needed to get back to her navigation station. Aldo Vandergamma, her supervisor, would be getting worried by now, wondering where she was. The reprogramming of Charlie should not have taken this long. She was hesitant to contact him on her comms, however, lest he ask her where she was and force her to lie. She had never lied to him before, and she wasn't going to start now.

Devory agreed to go with her, despite being reluctant to leave Mia alone. But Charlie assured him he would take care of her in his absence. Kendricks wouldn't know for sure that he and Anna helped Mia escape. The absence of Charlie from the Archive would automatically indict him, so that sealed the decision.

Vesta told them they could connect to the Recycling Sector through its telephone system, essentially a telegraph which the Recyclers had secretly wired up to tap into the primary communication system. If a message needed to be sent, they could use channel 81g, which would automatically perform end-to-end encryption so neither party would be compromised. The beauty of the system was that if Security or any other "official" stumbled onto that channel, all they would hear was traditional music sanctioned by the Continuance: melodic string arrangements that put even the most suspicious guard to sleep after a few minutes.

Devory and Anna thanked Vesta and headed back into the service tunnels while Mia and Charlie watched from a catwalk overlooking the "Mosh Pit" of the recycler village as Vesta had called it. Devory and Mia waved to each other before the hatch closed behind him.

Alone with Charlie, Mia slowly treaded the catwalk back to Vesta's quarters. The Mosh Pit, with its throng of workers moving about, was an assault on her senses, which, since her confinement in the Brig, seemed to have become hyper-sensitive. Workers bartered, haggled, and broke down everything from old pots and pans to dead LED lights to worn-out miners' equipment. The coppery tang of welding fumes mingled with the earthy smell of composted organic waste. Light glinted off metallic objects being sorted into massive bins, creating a constellation of flashing points below her. The constant percussion of metal striking metal formed a chaotic rhythm beneath the din of a hundred conversations. She wanted peace and quiet; the cacophony in her head was rapidly becoming too much.

The voices were coming back, stronger and more clamoring. Each one shouted a different argument, a discordant symphony of ancestral demands tearing at Mia's mind. Seneca Ramujan urged resilience, her voice a steady chant to rebuild trust, like the gentle but insistent lapping of waves against a shore. Eric, her father, bellowed warnings about the reactor's safety, his insistence that proper maintenance schedules not be skipped echoing with the authoritative timbre she remembered from childhood. A nameless Arkonaut from the Long Night screamed of betrayal, while a child's wail, lost and terrified, begged for salvation. The child's voice pierced her the deepest, making her chest constrict with an empathy so raw it felt like physical pain. She gripped the catwalk's rusted railing, her vision blurring as her mind imploded inward.

Charlie's titanium hand steadied her arm, his amber optics flickering with urgency. "Mia, your neural patterns are erratic. Heart rate 120, cortisol spiking. The voices—are they worsening?" His synthetic voice, calm and precise, pierced the din like a beacon.

Mia's breath caught in her throat, her knuckles whitening on the railing. "They're ... fighting, Charlie. Seneca, my father, others—they want different things. Save the ship, stop Prya, protect the reactor. I can't ... I can't think." Her knees buckled. Vesta, who had been watching Mia like a mother protecting her brood, rushed forward and caught her in her arms. "You're coming with me," she said, and guided Mia back to her quarters with Charlie's help. She sat Mia down on the dining room bench and put a blanket over her shoulders.

"I've seen this before," Vesta said, "Tunnels break 'em down, make their heads scream."

Charlie said, "I'm afraid it's much worse than that."

Vesta stared at Charlie with a pained expression. "Surely you're not saying it's that stupid rock they discovered, the one everyone seems to be going mad about?"

"That's the one," Charlie said, dead serious.

Vesta said, "I still think it's just a case of tunnel fever."

"Whatever it is," Charlie said, "we have to bring her back. Would you mind making her a cup of tea or something? She could do with a soothing drink."

"Well what do you know," Vesta said, getting up. "The can of nuts and bolts is telling me what to do!"

"It would be much appreciated," Charlie said, maintaining a diplomatic tone.

"You're lucky I agree with you," Vesta said, "otherwise I'd be out there auctioning you off to the highest bidder."

"I'm more valuable to you here, I promise," Charlie said. He turned to Mia and clanked his hands in front of her face. "Mia," he said, "can you hear me?"

Mia stared at Charlie, her eyes a thousand miles away.

Looking into her eyes for some kind of recognition, he said,

"Mia, there's something we promised each other. Do you remember?" He tapped his chest plate. "It's stored in here. Your failsafe."

At the word "failsafe," Mia's face seemed to orient, like a radio antenna searching for a signal.

"That's right, Mia ... your failsafe ..." Charlie's tone was encouraging now, coaxing her back to herself.

A flicker of recognition crossed her face. He had her full attention now.

"You can do it," Charlie said, beckoning her. "I am Mia Hawthorn, repeat after me ..."

This last prompt seemed to work and Mia began speaking: "I am Mia Hawthorn, scientist, daughter of Eric Hawthorn, Hero to the GenStar. This is a recording of my four highest values ..."

"Again."

Mia's eyes fluttered into focus as she said it, her voice stronger than before, "I am Mia Hawthorn, scientist, daughter of Eric Hawthorn, Hero to the GenStar. This is a recording of my four highest values ... the first ... the first is ..." She faltered, reaching for words that seemed just beyond her grasp.

"Consciousness," Charlie said, his voice soft but firm.

"Consciousness," Mia repeated, the word seeming to bring clarity with it, like a fog lifting.

Vesta, who was pouring a cup of tea while listening, didn't realize it was overflowing. She was completely captivated by the simulant's behavior. She cursed and straightened up the pot.

"Why?" Charlie asked. His eyes flashed green, his version of encouragement.

"Because its loss would be the greatest loss for the universe. It would no longer have a mirror to hold up to itself." Mia's voice grew stronger with each word, as if the statement itself was an anchor.

"Correct." Charlie's approval was like a tether, pulling her further from the edge.

Vesta came over with the tea. "Because it only happened once and we are the carriers of the torch," she said. She sat next to Mia and gently handed her the cup. The steam rose between them like a veil. "At least that's what my grandfather used to say. He was Chief Recycler, the best kind of philosopher."

Mia took the cup automatically and held it without drinking, the warmth seeping into her palms, another anchor to reality.

Charlie said, "Your second value, Mia, what is it?" His posture was attentive, focused entirely on her recovery.

"Fantasy," Mia replied, the word emerging with surprising strength.

"Very good," Charlie said. "Now tell me why."

"Because fantasy is what flows out of consciousness unbidden. It is the breath of consciousness as nectar is the gift of flowers." As she spoke, her eyes began to regain their usual intelligence, the cloudiness receding.

"And your third value, Mia?" Charlie leaned slightly closer, his sensors registering improvements in her vital signs.

Mia hesitated, her eyes searching inwardly, sifting through the competing voices to find her own.

Charlie and Vesta waited patiently. The only sound was the occasional drip from the kitchenette faucet and the faint raucous of the Mosh Pit's activity through the walls.

At last Mia said, "Science—because it is the great filter." Her voice had regained its characteristic authority, the voice of the researcher who had once commanded respect throughout GenStar.

"Too right!" Vesta exclaimed, slapping her knee. "It separates fantasy from fact. Not enough of that around here." Her enthusiasm was infectious, bringing a ghost of a smile to Mia's lips.

"And your last value?" Charlie asked quietly, his amber eyes steady on hers.

Mia's eyes had become clear, and she was picking up every little motion of Charlie's face, from the minute adjustments of his optical lenses to the subtle shifts of his facial movements. "Evolution, Charlie. Without an understanding of evolution, nothing makes sense. People just project their fantasies into reality when they ignore evolution."

"Well done, Mia!" Charlie said, flashing his eyes bright green, then yellow—his version of applause.

Vesta said, "I'm impressed. That's the fastest cure of tunnel fever I've ever seen. Sure as hell looks like it beats a hangover. That's our version of it. Get you drunk as a skunk. Make you regret you ever needed it in the first place." She laughed at the silliness of it.

Mia realized she was holding a cup of tea and said, "Is this for me?" She looked down at it as if seeing it for the first time.

"I promise it's free of ethanol," Vesta said, making a sign with her fingers over her heart. "Just chamomile and a bit of lavender from my private stash. Good for the nerves."

Mia took a tentative sip and, finding it to be just tea with an unusual but pleasant earthy aftertaste, gulped the whole thing down. Her body suddenly reminded her how dehydrated she was.

Vesta took the cup from her and said, "You must be thirsty." Her eyes scanned Mia with motherly concern. "Got more where that came from."

"I am." Mia's voice was steadier now, almost back to normal. The voices hadn't disappeared entirely, but they had receded to a manageable murmur rather than a roar.

Charlie leaned in again and said, "Mia."

"Yes, Charlie?" She turned to him, really seeing him now.

"What do we do now?"

Mia thought for a moment, her scientist's mind beginning to work again, hypothesizing and analyzing. "I don't know, Charlie. The failsafe seems to have worked—for now. I still don't feel one hundred percent, but at least I feel like I have something to combat it. Maybe this won't be the last time we have to do this." She rubbed her temples, feeling the lingering echo of the voices' presence.

"That may be true," Charlie said, his voice carrying a new gravity, "but what I meant was: what are we going to do with the rest of the GenStar? All the others that are affected by the Shard, who don't have a failsafe to bring them back?" The question hung in the air between them, weighty with implications.

This question seemed much harder to answer, and Mia accepted another cup of tea from Vesta, sipping it a few times before answering. The warm liquid seemed to clear her thoughts further, giving her the strength to confront the enormity of their situation. "I suppose ..." she said, "there's not much we can do. It seems like—"

A muted beeping sound drew here away from her thoughts, an urgent staccato that immediately set her nerves on edge.

Vesta immediately stood up and went over to a touch screen half-hidden behind a tapestry of woven wire, and tapped it, turning off the alarm. The screen flickered to life, casting a blue glow across her weathered features.

Charlie and Mia turned their heads and looked, tension radiating from both human and simulant.

It was Anna.

Her voice emerged, strained but clear, cutting through the soft background music that masked their communication. "Mia, Charlie, Vesta—listen. I have some urgent information." She paused, and they could hear her taking a steadying breath. "Carson has relieved Aldo of his navigation duties."

Mia looked at Anna's visage in the screen in puzzlement. "Why would he do that?"

"Because he wants to order a 10-megaton course correction."

"Okay," Mia said, unsure if she was following.

"It's not okay," Anna said, almost on the verge of tears. "He wants to steer GenStar into 82 Eridani."

Mia stared in disbelief, her restored clarity making the horror of this news even more acute. "Are you sure?" she said, feeling her mouth go dry despite the tea. The implication was staggering. 82 Eridani was a star, not their destination. It meant destruction, not arrival.

"Also just came back with a meeting with him. He couldn't talk him out of it." Anna's face on the screen was pale, her usually neat hair disheveled as if she'd been running her hands through it repeatedly. "According to Aldo, Carson was ... different. His eyes were glassy, and he kept talking about 'transcendence' and 'the ultimate merge.'"

"He's gone mad!" Mia heard herself say, the scientist in her re-coiling at the fundamental violation of their mission. The GenStar was humanity's ark, carrying the seeds of Earth to a new home, not a vessel for suicide.

Her mind raced through possibilities, each more desperate than the last. "Charlie, could we access the navigation controls remotely?"

"Aldo and I thought of that," Anna said, "The problem is Vasily could over-write the inputs."

"You're saying Vasily is in on this too?"

Anna shook her head. "We don't know what state of mind he's in. He's not answering Aldo's calls. All we know is Carson super-seded Aldo and gave his order directly to Vasily."

Mia abruptly stood up. "Then we need to send someone down

to propulsion right away." She looked at Charlie, then Vesta with urgency.

Vesta said, "There's no easy way to propulsion from here. You have to go through pneumatics. There's a service tunnel that runs along each piston. If they're preparing for a shot they'll lock them out."

Anna said, "True. But I might be able to disengage one of the blast doors for you. Can you stay in contact?"

"I don't have my comms with me," Mia said. "Kendricks took it from me."

Vesta said, "I wouldn't worry about that. We have some reverse engineered units that our raiding parties use."

"Thank-you," Mia said, "but—"

Charlie said, "I think she is saying she is coming with us."

"Are you sure?" Mia said, the worry returning. "We don't want any trouble. We just need to find out if Vasily intends to carry out Carson's order."

Vesta said, "The fact that Vasily is not returning Aldo's calls is clear sign that something is not right. Wouldn't you agree?"

Mia looked down, thinking. After a moment, she looked up again and said—to everyone: "Okay, then there's no time to waste."

– 25 –
DESPERATE MEASURES

Vesta chose two of her best raiders, the young men Dougnac and Jaffe. Dougnac, lean and wiry, had the sinewy grace of a tunnel scavenger, his face marked by weld scars. His makeshift crossbow hung across his back. Jaffe, stockier with broad shoulders, bore the Arkonaut's constellation tattoos and carried a handmade airgun loaded with darts.

"Tunnel 5B is your path," Vesta said, handing Mia a reverse-engineered comm unit, its casing scratched but functional. "It runs along the pneumatic piston. Tight, but Dougnac and Jaffe could navigate it blind." Her scarred hands adjusted the comm, her keen eyes softening with an unexpected tenderness that caught Mia off guard. "You ready for this, girl?"

Mia nodded slowly, drawing a well of strength from deep inside herself. "Carson's order will kill us all," she said, her voice steadier than she felt. "If Vasily's involved, we stop him." She glanced at Charlie, his titanium frame steady despite the dents and scrapes from their escape, then at the raiders, their weapons a silent promise of protection that simultaneously filled her with gratitude and dread.

Charlie's amber optics flickered, processing complex calculations with machine precision. "82 Eridani is a G-type main-sequence star with a surface temperature of approximately 5,600 Kelvin, not a habitable planet. Carson's course correction means complete incineration of the *GenStar* and all two-and-a-half thousand souls aboard. We must act now."

The bluntness of his assessment sent ice through Mia's veins. Two-and-a-half thousand lives—humanity's grand experiment—hurtling toward extinction because of the Shard's influence on Carson. The weight of responsibility pressed down on her chest, threatening to crush her resolve.

Vesta nodded, the scars on her face pulling tight. "Anna's working on the blast door codes from Navigation. Move." She led them to a rusted hatch hidden behind salvaged bins, their contents meticulously organized despite the apparent chaos—a testament to the Recyclers' ingenuity.

Dougnac pried the hatch open with a crowbar, the metal groaning in protest. The catacombs yawned before them, polygrout walls slick with decades of condensation, lit by flickering fungal strips that cast eerie shadows. Oil and metal choked the air, the distant chug of the pneumatic pressurization vibrating through the floor plates, rattling Mia's teeth.

The entrance gaped like a wound in *GenStar*'s infrastructure, its passageway used only by maintenance drones and desperate Recyclers. A memory flashed through Mia's mind: her father telling stories of these tunnels, how they connected the ship's vital systems like veins in a body, invisible but essential.

Dougnac took point, crossbow slung low, his lean frame slipping through the gloom with practiced ease. Jaffe followed, airgun ready, his staff tapping methodically for weak panels, comm unit crackling with occasional bursts of static. Mia and Charlie trailed, her boots slipping on accumulated grease, his sensors scanning continuously for threats or structural weaknesses.

The service tunnel, barely wide enough for Jaffe's shoulders, snaked alongside a massive pneumatic piston, its rhythmic groans echoing off the walls. The space felt alive, breathing with the ship's mechanical lungs. Rust flakes dropped from their scrapes against ancient surfaces, but she pushed on, her scientist's mind racing through possibilities, contingencies, fears.

"Stay low," Dougnac whispered harshly, securing a loose grate with belt wire, his eyes scanning the shadows. "Piston's live. One misstep, you're pulp." The warning wasn't theatrical. The tunnels claimed lives regularly, compressing careless scavengers to death with their shockwaves.

Jaffe's eyes flicked to a junction ahead, where three tunnels converged. "Prya's acolytes mark these walls with trapezoids now. No Security patrols, but watch it." His hand tightened around his airgun's grip.

Mia's breath caught at the mention of trapezoids—Cory Dunbar's obsession, now a cult symbol. The Shard's influence spreading like a virus through the ship's social structure. She remembered Cory's quarters, walls covered with the geometric shape, his son Jacob a victim of their incantations.

"Keep going," she said, pushing the memory away, the tunnel's weight pressing in from all sides. The walls seemed to contract with each breath, the air growing thicker, heavier. Claustrophobia clawed at her consciousness, but Charlie's Failsafe held firm, anchoring her in the present moment.

After what felt like an eternity, the tunnel widened slightly, a grated vent glowing with the Propulsion Sector's harsh industrial light. Dougnac pried it open with practiced efficiency, fingers working the rusted hinges without a sound, and they dropped one by one into a cavernous chamber where coolant pipes snaked overhead, hissing pressurized steam.

The space was heavily reinforced, more than any other on the ship, to withstand the enormous shockwaves from each delta-v blast. But the gantries were empty; technicians were nowhere in sight. It didn't make sense. But then a desperate clanging echoed from across the room. Their eyes darted to the sealed Bomb Room door, drawing their attention immediately.

Seong-Min Park, hectic and fierce, was hammering a heavy

wrench against the Bomb Room's reinforced door, her dark eyes wild with desperation. Sweat streaked her face; her jumpsuit torn at the shoulder. She spun as Mia approached, wrench raised defensively, then recognition flashed across her features.

"Hawthorn!" she gasped, lowering the tool slightly. "They said you were in the Brig!" Her gaze darted to Charlie, then to Vesta's raiders. "Timmerman's lost it completely!" The words tumbled out, panic-edged. "He beat Vasily half to death with a coolant pipe in there. Took his firing key. He's locked himself inside!"

Mia rushed to the door's small glass porthole, pressing her face against the thick shatter-proof glass to peer into the Bomb Room. The scene inside sent her heart plummeting. Vasily Petrov lay crumpled on the floor, blood pooling from his bald head, his burly frame unnaturally still. Lionel Timmerman, menacing and disheveled, clutched a firing key in white-knuckled fingers, its security lanyard dangling uselessly. His graying hair was matted with sweat, his eyes feverish as he muttered to himself, pacing before a bomb console with manic energy.

"Lionel!" Mia shouted through the porthole, her voice muffled but sharp enough to carry. "What are you doing? Open the door!"

Timmerman's head snapped up, his gaze locking on her with unsettling intensity. He seemed to focus for a moment, but it was clouded, distorted by something else—the same zealous glint she'd seen in Devory's eyes, in Chen's, in Prya's. The Shard's influence was warping his thoughts.

"Mia, you don't understand," he called, his voice trembling with

Shard-driven fixation. "I've heard Prya's sermons. The Shard's our destiny. But she's wrong! It's not about control—it's about merging, becoming one with the Shard itself. This bomb will vaporize *GenStar*, set us all free from our biological prisons!"

Mia's stomach churned with horror, her Failsafe's scientific rationality recoiling against such madness. The Shard wasn't just distorting thought patterns anymore. It was creating entirely new delusions, divergent interpretations that splintered the crew into frenzied psychopaths.

"That's insanity, Lionel!" she pleaded, pressing her palms against the door. "You'll kill us all! Open the door, let's talk about this!"

"No!" Timmerman shrieked, clutching the key to his chest like a holy relic. "The Shard demands sacrifice!" He turned toward the console—fingers trembling as they approached the activation sequence.

Seong-Min slammed her wrench against the door again, the metal ringing but remaining unyielding. "It's too solid!" she cried, desperation cracking her voice. "Nuclear safety protocols ... the door won't give. He's going to detonate it!"

Charlie's optics flared red, his neural net whirring frantically as he processed the situation. "The firing key initiates the detonation sequence," he stated, artificial voice tight with urgency. "We need that key, Mia, or he'll destroy everything we've fought to preserve."

Dougnac and Jaffe flanked the door, crossbow and airgun raised, their eyes evaluating the barrier with professional assessment.

"We can't shoot through," Dougnac said, his voice low. "Glass is blast-proof composite. Would take a shaped charge to breach it."

Before Mia could respond, heavy boots clanged on the gantry behind them. She turned to see Lora Vasquez storming toward them, her stocky frame bristling with weapons. A rifle gleamed in her hands, barrel sweeping across them with deadly precision. Her face, glistening with sweat, twisted with fury.

"Hawthorn, you're done running!" she declared, weapon trained steadily on Mia's chest. "And you brought Recycler scum with you. Drop your weapons, all of you!"

Dougnac's crossbow snapped up reflexively, Jaffe's airgun trained on Vasquez without hesitation.

Mia instinctively backed behind Charlie. "I haven't done anything wrong," Mia declared.

"Wrong?" Vasquez parroted. "Everything about you is wrong! And now that you've teamed up with these jokers, you've really outdone yourself. You've proved that you were feral all along. Kendricks was right to mistrust you. I'm not going to say it again. All of you, put your weapons down!"

Dougnac said, "It's two of us to one of you. You want to do this?"

Vasquez scoffed. "Bullets fly faster than your home-made toys. I'm ready when you are."

Mia's eyes shifted. It was Seong-Min Park. She had crept up behind Vasquez, the wrench raised, ready to strike.

Mia shook her head. She didn't want to see bloodshed.

Vasquez observed Mia's reaction. But it was too late. Just as she spun around, the wrench came down. It didn't knock her out, but it was enough to make her crumple to the floor like a rag doll.

Jaffe swiftly closed the distance and took Vasquez's rifle out of her hand.

Vasquez, dazed, but still conscious, mumbled something.

Charlie said to Dougnac and Jaffe. "Thank you. Mia and I can handle this by ourselves from here. You go back to your people. Warn them."

"No," Jaffe said. "We'll stay here and see if we can breach Bomb room and stop Lionel. We have what we need now." She indicated Vasquez's rifle in hand.

Charlie nodded. "Seong-Min. Maybe you should go and find Phil?"

Seong-Min agreed.

"Go!" Dougnac said. "Go and see if you can stop the Captain."

Charlie and Mia didn't have to be convinced. They eased themselves through the narrow gap behind the cooling pipes and slipped into the ventilation tunnel that fed them.

It quickly grew dark but Charlie used his infrared vision to guide the way. He led Mia through the ventilation duct, taking crucial turns at cryptic junctions, each step taking them further from the chaos they left behind.

As they went, Mia began thinking about how much time they had left. She knew prepping a device took time, it wasn't just a matter of pressing a single button. It was common knowledge that procedure was required. But how long exactly she couldn't say under the current circumstances. Lionel, in his heightened state of mind, it could take longer, or shorter. At most probably an hour or less.

It struck her as incomprehensible. To think that everyone's lives would end in less than an hour if Lionel wasn't stopped. That their consciousness would be extinguished. It seemed like a cosmic crime. She hoped that Vesta's Raiders could break through, but that door—it wasn't the strongest on the *GenStar* for a no reason.

Charlie spoke up, breaking her thoughts. "I've done some calculations," he said.

"Oh?"

"I put the probability of saving the *GenStar* at less than twenty-two percent."

"What are you saying Charlie?"

"I'm saying it's time to make some painful decisions."

Mia sensed the implications right away. It was in the tone of the simulant's voice. The pitch was heightened, full of tension. "No Charlie," she said. "I don't think I can do that."

"What's the alternative?"

Mia didn't have an answer.

"I suggest we start thinking about our loved ones."

Mia almost melted. "Yes, Charlie. You're right. Let's save our loved ones."

– 26 –
NO TIME FOR MIRACLES

Mia followed Charlie through the darkened access tunnel, her breath shallow, boots clanking on steel as they headed to Mining Control. The steady chug of the pneumatic pumps faded as they moved away from the Propulsion sector. Her mind burned with what she had seen: Vasily lying in a pool of blood, Timmerman's cold, inhuman eyes, and the green glow reflected in his face, evidence he'd accessed the timing panel, the countdown clock set by his Shard-corrupted brain.

"Just up ahead," Charlie said, pointing to the pressure door stenciled Mining Control Access – Authorized Personnel Only. His voice was steady, but his servos whirred faster than usual, the subtle increase in pitch a sign that he had switched to full power mode.

He punched in the door key code, relieved Anna hadn't reprogrammed him or wiped his security protocols.

To Mia's surprise, the control station was unoccupied. Or maybe she shouldn't have been surprised. Nothing was normal anymore. A flickering monitor drew her attention. She approached it and scrutinized the image. It was mostly darkness, with a pale glowing outline of a hunched figure. Mia figured it was a miner at work. She marveled that someone could still think of working when everything was falling apart.

"Have a look at this," Charlie said, distracting her. He pointed to a glowing orange panel light above a digital text display. The words read: Murchison: 19F.

"That's Devory!" Mia exclaimed. She looked at the video monitor again. She frantically searched for the comms panel.

"Here," Charlie said, pointing to a grilled speaker with a microphone above it.

Mia rushed over and buzzed the intercom button.

The figure on the monitor continued uninterrupted.

"Dev!" Mia cried into the mic. "It's me, Mia. Can you hear me?"

The figure momentarily stopped what it was doing, then continued again.

"Dev!" Mia repeated. "You're not hearing things. It's me. I'm up here in Control. You gotta listen to me."

Devory slowly turned around and in the dim light, Mia saw his features behind his helmet visor.

"That's right," Mia said. "You can hear me. You have to listen, please, I've got something important you need to know."

Devory stood still, staring at the tunnel camera.

"I love you, Devory. I really do. I understand now. You need to find more pieces of the Shard. I respect that. But something's come up. You have to listen to me now ..."

Devory seemed to waver. He looked at the camera, then away again, for whatever he was hunting.

"I've got something awful to tell you ..." She swallowed heavily. "Timmerman's killed Vasily. He's activated a bomb. We have 45 minutes, maybe less."

A sharp rasp of breath came through the comms. Mia didn't know what to make of it. Had their communication been cut?

She pressed the buzzer again, and leaned closer into the mic. "Did you hear me? Timmerman has started the timer on a nuclear device. We don't have much time ..."

Devory slapped his helmet. The speaker squealed and Devory's voice came through again, thin and high-pitched. "I'm hearing you—come again. Timmerman has what?"

"He's murdered Vasily and has commandeered a bomb."

"No way!"

Devory walked toward the tunnel camera and dropped out of vision. "I'm coming up," he said. "Wait for me."

"We don't have much time. Hurry!"

Charlie looked at Mia, his eyes glowing red. His warning sign.

"I know, Charlie," Mia said. "Every minute counts."

The elevator control panel lit up, showing the elevator was rising.

Mia began pacing. She couldn't contain herself.

Charlie said, "When he gets here, what are we going to do?"

"I don't know, Charlie. Dev's smart. He'll think of something."

The elevator control panel light reached the top level.

Across the floor and behind a bulkhead they could hear the elevator gate clanging open. Mia rushed to the door.

The door swung open and Devory burst in, his mining suit already doffed, his face matted with sweat, his eyes burning with intensity.

Mia ran to him and threw herself into his arms. "Oh Dev," she cried. "It's you! You heard me."

"I heard you," Devory murmured. He released her and looked into her eyes. "Say it again. I want to hear it."

"Timmerman—"

"No. The other thing."

"I love you."

"And I love you too," he said. It was pure and without embarrassment. A burly miner soft as a newborn puppy.

"Ah ... girls and boys," Charlie said, intervening.

"How much time have we got?" Devory asked, breaking free from Mia.

"I don't know," Charlie said. "We had maybe an hour when we left the Bomb Room. We might only have 45 minutes left. It's hard to say."

As if his words had been a premonition, a klaxon sounded, followed by an automated voice announcing, "All Arkonauts, a nuclear device has been activated. The countdown has started. Secure all loose items within 45 minutes. Next announcement in 15 minutes."

Devory and Mia looked at each other with terrified expressions.

"What will we do?" Mia asked frantically.

"Let me think," Devory said, walking over to the control panel, as if the answer lay there somewhere. He leaned forward and planted his calloused hands on the cold metal. "There's no stopping Timmerman," he said after a few moments. "We'll never reach the Bomb Room in time. The only option now is the Landing Craft."

"The Lander?" Mia said. "We don't even know if it's—"

"Rivel does," he said. "She's the only one who knows how to detach it. But we'll need a pilot."

Mia knew right away that there was only one choice: "Aldo."

Devory nodded grimly. "Exactly." He punched the intercom and called Rivel. There was no answer.

He spun around. "The Engineering Bay. Let's go!" he said.

They moved fast. Through access ladders, sub-hallways half-forgotten since early-phase excavation. The Engineering Bay loomed ahead, its orange doors splayed open like a hungry mouth.

They crossed the floor, hurrying past massive pieces of equipment—lathes, CNC machines, gantry cranes, and gas cylinders. Rivel's office was tucked away behind a row of racks holding tubing of various types.

Rivel sat cross-legged on the floor, surrounded by floating holo-diagrams—concentric fields, Shard geometries, electromagnetic lattices curled like fern fronds.

"Rivel," Mia called. "We need you."

The woman didn't look up. "Do you understand what this means? The Shard isn't a crystal—it's a syntax, the interface between awareness and field, solving the ancient problem of mind versus matter."

"Rivel!" Devory stepped forward, gripped her by the shoulders and shook her, gently but urgently. "There's no time. We need to launch the Landing Craft."

The mention of the "Ark" seemed to snap her out of her hypnotic trance.

"Landing Craft?" she said, looking up at them like a child who didn't quite understand.

"Yes," Devory said. "Mia, tell her."

"You heard the alarm," Mia said. "A bomb is set to go off. We have less than 30 minutes now to detonation."

"A bomb?" Rivel asked. "In God's name why?"

"Timmerman," Mia said. "He's gone mad and has killed Vasily. He's started the countdown and there's nothing we can do now except leave. Do you understand? We need to launch the Landing Craft."

"The Landing Craft," Rivel repeated, as if saying it would make it real.

"Yes!" Devory said, grabbing her, enticing her to get up and start moving.

Rivel stood up and looked at them, blinking. The fog slowly lifting from her mind. "I can get it loose," she said. "But I can't fly it."

"We know," Devory said. "Aldo will fly it."

"Aldo," Rivel nodded. "Yes. He can fly."

"We gotta go now," Mia said, tugging on Rivel's arm.

Rivel nodded, the situation fully dawning on her. "I'll need Aldo in the pilot's seat while I decouple ground support. Is he already there?"

"No," Devory said.

"In that case we had better call him now," Rivel said. She leaned her chin into her comms unit. "Aldo, this is Rivel. Respond."

There was nothing, just a static click.

"Aldo, respond!" Rivel repeated.

There was a pause, then another click. "Aldo here. What's going on? Why has a countdown been started without my authority?"

"There's no time to explain. We're going to launch the Landing Craft. Get your ass down there ASAP."

"What? You want to launch the Landing Craft?"

"Aldo! Listen. Timmerman has killed Vasily. He's going to blow the ship!"

There was a long silence. Then, quietly, with heavy resignation in his voice, Aldo said, "Okay, I'm on my way. But not without Anna."

"Isn't she with you?"

"No."

Rivel cursed under her breath. Then, "Aldo, whatever happens, we need you in the pilot's seat. Do you understand?"

"Yes, I understand."

"Then please, hurry!"

"I'm hurrying!"

Devory looked at his watch. "Cutting it fine."

Rivel led them to the Landing Bay. On the way she said, "All we have to do is untether it and move it away. We don't have to run the mains. Just the thrusters."

Devory said, "Last number I got was 83 percent LOX in the main tank. Has that increased at all since our last ice haul?"

"We're up to 87 percent," Rivel said, "give or take a couple, counting boil-off."

"Jeez, I hope Aldo can fly that bucket," Devory said, doing mental calculations on the numbers.

"He's got the highest number of sim hours of anyone on board," Rivel said. "If there's a trajectory, he can fly it."

"What if there's no trajectory?" Mia asked.

"I don't want to think about that right now," Rivel shot back. "Let's get off this rock first and then worry about that."

"Sure," Mia said.

They rounded a corner.

And stopped.

Kendricks stood like a sheriff guarding his town. Arms outstretched, feet apart, gun aimed squarely at their chests.

— 27 —
SACRIFICE

Kendricks' eyes glinted with a feverish intensity, the barrel of his pistol sweeping across the group. "Nobody moves," he growled, his voice low and jagged, like two rocks grinding together. "The Landing Craft is the people's property. Any attempt to unilaterally commandeer it will be regarded as treason."

Mia's heart pounded. She stood frozen beside Devory, Charlie, and Rivel, the air in the Landing Bay thick with the weight of imminent violence. Rivel's hand twitched, fingers brushing the utility knife clipped to her belt. Mia saw the movement and wanted to scream *No, don't!*, but her throat clamped shut.

"Kendricks," Devory said, stepping forward with hands raised. His voice was calm, deliberate, the tone of a man talking someone off a ledge. "Put that gun down. Come with us if you want. But don't stop us from trying to save ourselves."

Mia shot Devory a glance, as if to say *What are you doing*?

Kendricks sneered. "Save yourselves? From what? The Shard? You're deluded! I've got the situation under control."

"No you haven't," Mia said, speaking up, surprising herself. "Timmerman is going to start the countdown on a bomb. He intends the blow the ship."

Hearing this, Kendricks momentarily faltered. But he quickly recovered. "I don't believe you. You're lying."

"Why don't you ask your deputy?" Mia challenged.

Kendricks accepted. He activated his comms. Called Vasquez. After a few attempts, her voice came through. "Your timing is inconvenient as usual," she said. "I've been compromised."

"What do you mean 'compromised'?" Kendricks shot back.

"They've taken my rifle and they're trying to get into the Bomb Room. But it's futile."

"What do you mean it's futile?" Kendricks demanded angrily. "Use the over-ride code."

"I would have, if Timmerman hadn't short-circuited the door."

"What—what is he doing?"

"He appears to be working on a device."

Kendricks looked at Mia.

Mia said, "Don't say I didn't tell you."

Rivel stepped forward.

Kendricks switched his aim to her chest.

Rivel's eyes narrowed. Her fingers tightened around the knife's handle, her stance shifting forward.

Mia's stomach dropped. *She's going to try something.*

"Put the knife down, Rivel," Kendricks ordered, focusing his aim at her heart. "I see you reaching. Don't be stupid."

"You're the stupid one," Rivel shot back. "Timmerman's going to blow us all to hell, and you're playing the hero—for what? Your own piece of the Shard?"

Kendricks' finger twitched on the trigger. "Last warning."

It happened in a blur.

Rivel lunged. The knife flashed in her hand, aimed for Kendricks' arm. She was fast, driven by desperation. But the pistol was faster. A deafening crack rang out.

Rivel staggered. A bloody patch appeared on her chest. She dropped to her knees, the knife clattering beside her, her breath coming in shallow gasps.

"No!" Mia cried, lunging forward, but Devory yanked her back.

Kendricks swung the pistol toward them, face twisted in rage. "Anyone else want to try me?"

Devory's fists clenched. Yet he didn't move. Mia felt the tension in his grip, the coiled energy of a man biding his moment.

Charlie shifted slightly, positioning himself between Kendricks and Mia, servos changing in pitch.

Mia knelt and comforted Rivel. "Why?" she asked, looking up at Kendricks with ferocious venom.

"She asked for it," Kendricks said, his tone scathing.

Rivel coughed, blood flecking her lips. "We're even now," she said to Mia.

"Shhh," Mia whispered. It was true. Mia's father had sacrificed himself for others, Rivel among them that day in the reactor room.

Rivel tried to raise her head, but lacked the strength. "Make him proud, Mia."

Mia wished for all the world she could do something. But all she could do was watch Rivel fade.

As a last gesture, Rivel beckoned her closer.

Mia put her ear to Rivel's lips.

Rivel gulped a breath. "Why ..."

"Why what?"

"Why do we feel everything is ... connected?"

Mia stared dumbly at Rivel. She had no answer.

Rivel exhaled.

Mia shook her.

But Rivel's head slumped forward and went still.

For a moment, Mia couldn't move. It was like someone had drugged her with a paralyzing agent. Stopped time.

Not until a hand touched her shoulder did she realize her body was still physically hers, and she spun around viciously.

It was Devory. His face was full of anguish.

Behind him, Kendricks was unmoved.

"You bastard!" Mia cried. "Look what you did!"

Kendricks merely shrugged—as if to say, *it's not my problem.*

Mia brushed some dust from Rivel's cheek, as her vision blurred. "I'm so sorry," she murmured. "I'm so sorry."

Devory was slowly shaking his head, seething at Kendricks.

Kendricks said, "You want it? Come on, get it."

Mia saw Rivel's knife lying close by. While Devory distracted Kendricks, she hooked it with her foot and dragged it closer until she could reach it with her hand.

Devory held his ground, but had raised his fist, shaking it at Kendricks. "You'll pay for this Kendricks," he said. "One way or the other."

Kendricks laughed callously.

In that moment, Mia threw the knife at him. It was more an angry gesture than an effective attack. It grazed his neck, then peeled away.

Kendricks expressed surprise, put a hand to his neck.

That was all it took.

Devory moved—a blur of motion, launching at Kendricks like a rugby player fielding a tackle. Kendricks made an attempt to bring the pistol around, but Devory closed the distance in two strides. The shot went wide, grazing Devory's shoulder and tearing his jacket. Devory roared, slamming into Kendricks with his full weight, targeting the man's midriff.

The pistol skittered across the floor.

Kendricks fought back, fists hammering Devory's ribs. But Devory was relentless. He pinned him down, then held his elbow at Kendricks' throat.

"You can't stop it!" Kendricks wheezed. "The Shard—"

"You're done," Devory declared.

Mia noticed the same handcuffs dangling from Kendricks' belt that he had taunted her with during their Master Airlock encounter. It gave her an idea.

"Roll him over," she said, instructing Devory.

Devory understood.

Kendricks tried to resist, but Devory grabbed one of Kendricks' arms and yanked it back hard, making Kendricks comply.

Mia unclipped the cuffs from Kendricks' belt and slapped one of them onto the arm Devory had pinned down.

"Over there," Mia said, indicating a sturdy coolant pipe.

Devory agreed. He hauled Kendricks like a sack of potatoes and together, he and Mia clipped the other cuff to the pipe.

"The key," Mia demanded.

Kendricks merely grinned like an idiot.

"It's in his pocket," Devory said.

Mia fished it out and threw it down the tunnel as far as she could.

Kendricks slumped, realizing he had played his hand.

A klaxon blared, sharp and metallic.

"Thirty minutes to detonation," the automated voice intoned. "Fix all loose items. Secure children into their blast seats."

Mia couldn't believe the automated system was issuing advice for a doomed mission.

Kendricks laughed hoarsely. "You'll never make it. The Landing Craft can't be launched in less than thirty minutes. Rivel was the only one who knew the complicated release sequence. Carson made sure of it—one person, one key."

Devory scowled at him. "At least we'll die trying, you piece of shit!"

He turned to Mia, his eyes wild with adrenaline from the hand-to-hand combat with the Security Chief. Blood seeped through his torn jacket, but the wound appeared superficial.

Mia went up to him and put her hands on Devory's chest. "The uncoupling sequence," she whispered, her eyes fixing on his. "Rivel was the only one who knew it." Her gaze drifted to the engineer's body, lifeless on the floor. Another sacrifice in the long chain of deaths stretching back to her father. She had no qualms about leaving Kendricks to his fate—she should have—but Rivel's death pushed her over the edge. An animal instinct drove her now. The time for moralizing was over.

Devory was cursing under his breath. He turned to Charlie. "Do you have any schematics—anything that could help us?"

"I have schematics," Charlie replied, his optical sensors dimming momentarily as he accessed his memory banks. "Accessing them now. I have partial information on the Landing Craft coupling mechanisms, but the final release requires a six-digit command key known only to Rivel."

Mia said, "Can we do this while moving? We have to hurry."

"Absolutely," Devory said, already striding toward the exit.

They left Kendricks and proceeded to the Landing Craft Bay doors, one more turn around the next corner. It was a large junction, specially designed for access from three directions, a testament to the careful planning of their ancestors sixteen millennia ago. As they left, Kendricks shouted after them, his voice cracking with desperation.

"What about me? You can't leave me here!"

They ignored his cries and rushed to the doors.

Just as they arrived, Aldo turned up. And to Mia's surprise, Anna was with him.

"Boy, are we glad to see you guys," Mia said.

"Same here," Anna said.

"Where are we up to?" Aldo asked, the urgency in his voice evident.

"If you can climb aboard, I'll take care of the hold-down clamps," Devory said.

"Okay," Aldo replied.

Charlie was working on the entrance doors. "Got it!" he said.

The doors slid open to reveal the Landing Craft, a lift-body-shaped vehicle with a bullet nose and manta ray wings. The heat tiles covering its belly were made from sintered regolith, and its upper body from recycled aluminum obtained from Arkonaut bed bunks and chairs—just one of the many sacrifices the Arkonauts had made to ensure their promised future.

The hold-down clamps needed to be released only after the outer hatch was opened. Aldo reminded Devory that this was the correct procedure. As soon as the clamps were released, the centrifugal force created by the *GenStar*'s rotation would automatically fling the spacecraft away from the ship. Thrusters could be used to maintain orientation, but it wasn't necessary. The main challenge was chilling down the main engines. They could not be fired until this was accomplished. If they needed to get away from the ship in a hurry, then they needed to get that process underway immediately.

Mia stared at the vessel in trepidation, having never conceived of actually flying in it or being part of the first crew. The first flight was slated only to take an experienced flight team and stay in close orbit around the ship. Only after such tests had been completed was it to re-dock and load up the selected "landing party"—about 120 Arkonauts drawn from a ballot. Most of these would be engineers, agriculturalists, and scientists, as they had to not only survive on Eridani-B but build two new crucial pieces of technology: an in situ fuel production plant, and a booster rocket capable of sending the Landing Craft back into orbit to rendezvous with the *GenStar* to bring a new batch of Arkonauts down to Eridani-B. Everyone knew this would take years. It was a long-term plan. Short, perhaps in terms of the big scheme, but those left on board didn't want to wait any longer than necessary. It was a plan of hope.

Aldo and Anna climbed up the scaffolding that still surrounded the vessel and made their way into its interior. The craft would have to break away from the scaffolding on release. There was no time to remove it.

Devory looked at Mia and urged her to climb up ahead of him.

"Who is going to release the hold-down clamps?" Mia asked.

They both knew who it would be. They looked at Charlie—the only one capable of doing the job and surviving the vacuum of space.

"I will attach a tether to myself," Charlie said. "You can bring me onboard once we are safely away."

Mia listened, but a sense of unease diverted her attention. Something wasn't right. It was all happening too fast.

"What's wrong?" Devory asked.

Mia gave him a fretful look.

"What is it?"

She started backing away, speaking as she went. "There's one more thing I need to do."

Devory stared, confused.

Mia said, "I'll come back. I promise. Ready the craft. Go!" She ran back out of the bay.

"Where are you going?" Devory called after her. But it was too late. She was already gone.

– 28 –
THE WEIGHT OF TRUTH

Mia's boots pounded against the polygrout floor as she sprinted through the *GenStar*'s access tunnels, the klaxon's relentless wail in her ears a reminder that she was on a sinking ship.

Ten minutes, she thought, her mind a whirlwind of fear and purpose. She had to find Tiber Solis in the Archives. She couldn't leave him behind.

The ship shuddered beneath her feet—another failsafe system collapsing under Timmerman's sabotage. The corridor lights flickered in sync with the oscillating pressure in the central coolant system; she could almost feel the 40 Hz pulse of the Shard rippling through the ship's infrastructure. Life support readouts on wall panels she passed flashed emergency warnings—oxygen levels dropping precipitously in several sectors where blast doors had automatically sealed.

As she rounded a corner, the electrical workshop came into view, its door ajar, sparks flickering from a half-dismantled panel. Callum sat amidst a pile of tangled wires and circuit boards, his face etched with worry, a comms unit pressed to his ear.

"Rivel, come in. Where are you?" His voice was hoarse, desperate. He seemed to be oblivious to the countdown, but this was no time to hesitate.

Mia skidded to a stop, her breath tight in her chest. She knew the truth would crush him—Rivel's lifeless body, the blood pooling beneath her, the way her remaining eye had fixed on some distant point as life slipped away. But there was no time for grief, not now. She steeled herself, forcing her voice to stay steady.

"Callum, Rivel's at the Landing Craft. Hurry and go join her there."

Callum's head snapped up, his eyes wide with hope. "She's there? You're sure?"

"Yes," Mia lied, the words bitter on her tongue. "Go now. She needs you."

She hated herself for it, but saving his life meant more than sparing his heart. Rivel would've wanted that. Her final words a puzzle she promised herself she would share with Callum—once he was safe, once his consciousness was secured. He would know what she meant. She had to believe that.

Callum scrambled to his feet, knocking over a tray of chip

components. "I'm going. Thanks, Mia." He paused, then looked at her, suspicion clouding his features. "You're heading there too, right?"

"Yeah, but I need to grab something first." She hesitated, then pointed to his comms unit. "Can I borrow that? You won't need it if you go straight to the Landing Craft, but I might need to let Devory know I'm okay so he doesn't worry."

Callum seemed to accept that; it made sense in the heat of the moment. "Fine. Just ... be quick."

He handed her the comms unit, its casing warm from his grip, and took off toward the Landing Bay without another word. The device showed a full signal—unusual, given the general communications interference that had plagued the ship since the Shard's discovery. Trust Callum to have probably figured out a workaround.

Mia clutched the device and ran, her legs burning as she navigated the maze of tunnels toward the Archives. Fortunately, she knew this part of the ship. She could make good time, so long as she didn't stop—or falter.

On the way, she made a short detour to her private quarters and grabbed her father's medal. Deep down, she knew she couldn't leave without it, even though it was too late to find her mother. It was either Tiber or her, and she made her decision.

The Archives loomed ahead, its heavy doors splayed open from the recent attack. She burst inside, her eyes adjusting to the dim light. The emergency power cells cast everything in a pale blue glow, highlighting the devastation. The room was a wreck—but that's not what her eye searched for.

In the center, Tiber Solis knelt amidst the destruction, his thin frame hunched over a pile of loose pages. He was carefully gathering them, his fingers trembling as he tried to reassemble a book, its cover embossed with the *GenStar*'s crest. His face was serene, almost reverent, as if the world wasn't crumbling around him. A portable oxygen concentrator sat beside him, its soft hum almost lost beneath the distant klaxon.

"Tiber!" Mia called, her voice sharp with urgency. "Forget about it! There's no time!"

Tiber looked up, his eyes calm, almost distant. "Mia, it's you!"

"Yes, it's me. I've come here to help you."

Tiber chuckled humorlessly. "I know you love history, but not that much."

"No, Tiber. To the Landing Craft. Come on! We have to go now."

Tiber gave her a long, sad look. "No, Mia. My work is here."

She froze, confusion knotting her chest. "What are you talking about? Timmerman's set the timer on a bomb. It's going to go off in less than ten minutes! You have to come. It's our last chance!"

Tiber shook his head, a painful smile tugging at his lips. "I'm not going."

Mia stared at him, her mind reeling. "What? But ... why?"

He set the pages down, his movements slow, deliberate. "I'm dying, Mia. Cancer. I've been dying the whole time. I just never had the heart to tell you."

Her knees buckled, and she grabbed a nearby toppled shelf for support, the metal cold beneath her fingers. "No. That's not—surely we can do something! Operate, treat it, anything! Just come with me to the Landing Craft!"

Tiber's eyes softened, but his voice was firm. "No. It's inoperable. Ask Dr. Chen. She'll explain. Something to do with my younger, more promiscuous days."

He gestured to the historical texts surrounding them. "Besides, someone needs to stay with the knowledge we couldn't save."

Mia's hands shook as she fumbled with Callum's comms unit, her fingers clumsy with panic. "Lilien ... Lilien, this is Mia. Please respond. Lilien!"

There was a faint crackle, then Dr. Chen's voice came through. "Mia? Is that you?"

"Yes. I'm borrowing Callum's comms unit. It's me."

"I'm glad it's you," Chen said. "I've been looking for the Shard fragment the whole time. I don't understand. It was here in my lab."

"About the fragment ... I can explain."

"Please do."

"First I need you to tell me something ... about Tiber. Is it true he's dying of cancer?"

There was a pause, then Chen's voice, clipped. "That's confidential information, Hawthorn. Medical ethics still apply—even if we're all about to die."

Mia's desperation boiled over. "I'll tell you where the Shard fragment is if you tell me about Tiber."

Another pause, longer this time. A distant explosion vibrated through the comms—another system failing.

"Fine. Yes, it's true. Tiber has inoperable cancer. Stage Four lymphatic metastasis. He's got months, maybe less. The nano-treatments only slow it down."

Tears welled in Mia's eyes, blurring her vision. She choked back a sob, her gaze locked on Tiber's frail form. "Thank you," she whispered into the comms. "The Shard fragment's with Charlie at the Landing Craft. Go there now. I'll meet you there."

She lowered the comms, her voice breaking. "Why didn't you tell me?"

Tiber rose, his movements slow, deliberate. "I didn't want to burden you. You've carried enough."

He reached for the lanyard around his neck, lifting it off. The data stick dangling from it gleamed in the faint light, engraved with the *GenStar*'s insignia.

"This holds everything—my records, the voices of our ancestors, the history of the *GenStar*. Every genetic sequence, every engineering specification, every cultural artifact I could salvage. You're the future, Mia. Take it with you. Tell your children, and your children's children."

He stepped forward, slipping the lanyard over her head. The weight of the data stick settled against her chest, heavy with responsibility. The stick's memory capacity was staggering—exabytes of compressed data, enough to rebuild their civilization if needed.

Mia's tears spilled over, and she threw her arms around him, hugging him tightly. "I can't leave you," she whispered.

"You have to," Tiber said, his voice gentle but unyielding. "I'm happy here. Believe me. There's nowhere else in the world I'd rather be."

Mia stared at him and knew it was true. "Goodbye, Tiber."

She pulled back, her heart breaking. His face was gaunt in the emergency lighting, but his eyes were clear—she let them be her last memory of him.

She turned and ran, the data stick bouncing against her chest. She stopped one more time at the entrance and looked back. She expected Tiber to be looking at her, waving goodbye—but he was already crouched down, collecting more papers, the outside world no longer part of his world.

Mia turned away again and sprinted through the corridors, activating the comms unit as she went, her voice ragged with urgency. "Devory, it's Mia. I'm running as fast as I can. Callum and Chen are on their way to the Landing Craft. Wait for us!"

Devory's voice crackled through, strained but relieved. "Hurry, Mia. We're almost ready, but time's running out. Charlie's preparing to open the outer hatch. Seven minutes left."

"I know," she said, pushing her legs harder, the Landing Bay her only focus.

The ship trembled again, more violently this time. A nearby

conduit burst, spraying coolant across her path. She leapt through the spray, the liquid instantly flashing to a cloud of vapor.

Tiber's sacrifice, Rivel's death, the weight of the *GenStar*'s history—they all drove her forward. Behind her, the Archives grew smaller, sixteen millennia of human achievement condensed into a dying man and a data stick.

She wouldn't let them down.

— 29 —
LAST SECONDS

Mia's lungs burned as she sprinted back toward the Landing Craft Bay, the data stick around her neck thumping against her chest with every stride. The *GenStar*'s corridors blurred past— flickering lights, hissing conduits, the klaxon's wail a relentless drumbeat counting down to oblivion. Tiber's serene face haunted her, his sacrifice a weight heavier than the exabytes of history she now carried. She pushed the grief aside. There was no time for it.

She rounded the junction, the Landing Craft Bay's massive doors looming ahead, when a voice—raw, hysterical—stopped her cold.

"Go on, do it! Put me out of my misery!"

Kendricks.

Mia skidded to a halt. She listened.

"You can't do it, can you?"

She backtracked, heart pounding, and peered around the corner into the access tunnel where they'd left him. The sight froze her blood.

Callum stood over Kendricks, who was still handcuffed to the pipe, his face a mask of defiant rage. Callum clutched Kendricks' pistol in both hands, the barrel aimed at the man's head. His fingers trembled violently, the gun wavering, his eyes wild with a mix of fury and fear. Kendricks' sneer dared him to pull the trigger.

Mia's breath caught. She rushed forward, her voice cutting through the tension. "Callum, don't do it! You'll be just like Kendricks—a cold-blooded murderer!"

Kendricks' head swiveled toward her, his eyes glinting with venom. "You hypocritical bitch! What do you think handcuffing me to this pipe is? A retirement present?"

Mia's jaw tightened, but she kept her gaze on Callum, her hand outstretched, palm open. "You're uncontrollable, Kendricks. You left me with no choice."

"Choice?" Kendricks shrieked, his laugh bitter and sharp. "You can choose to let me join you and your co-conspirators. We can all be one happy family then, what do you say to that?"

Callum's trembling intensified, the gun shaking so hard Mia feared it might go off accidentally. His face was pale, sweat beading on his brow. For an instant, her heart skipped a beat. He couldn't possibly know about the lie she told. He must have found

Rivel lying there and assumed Kendricks killed her in the interim.

"Callum, listen to me," Mia said, stepping closer, her voice steady despite the chaos. "He's not worth it. Don't let him drag you down with him."

Kendricks opened his mouth to retort, but before he could speak, there was a load bang. Mia flinched, her heart lurching as the shot echoed in the tunnel. But Callum's aim was off—wildly so. The bullet missed Kendricks' head by inches, punching into the tunnel wall with a spray of sparks and polygrout dust.

Kendricks flinched, his bravado cracking for a split second before his sneer returned. "Pathetic."

Mia seized the moment. She lunged forward, wrenching the pistol from Callum's shaking hands. "Come on," she said, her voice firm but gentle. "He's not worth it."

She turned toward the Landing Craft Bay, motioning for Callum to follow. But he didn't move. Instead, he dropped to his knees beside Rivel's body, still lying where they'd left her, the blood now congealed around her. His hands hovered over her, trembling, as if he could somehow bring her back.

"Callum, no," Mia said, her voice breaking. "Leave her here. We haven't got time."

He didn't respond, his fingers brushing Rivel's cheek, muttering words under his breath. Mia's chest tightened—she wanted to let him grieve, to give him the moment he deserved, but the clock was merciless.

Her comms unit crackled, Devory's voice urgent. "Mia, we've

got two minutes. It takes two minutes to open the outer hatches. It's now or never. Move!"

The words were a jolt. Mia grabbed Callum's arm, yanking him away from Rivel with a strength born of desperation. "She'd want you to live," she said, her voice fierce. "Don't let him win!"

Callum's eyes met hers, glassy with unshed tears. For a moment, she thought he'd resist, but then he stumbled to his feet, his movements sluggish, like a man waking from a nightmare. He followed her, half-running, half-staggering, as they raced toward the Landing Craft Bay.

The massive doors stood open, the bay a cavernous expanse dominated by the Landing Craft's sleek, aerodynamic form. Mia and Callum climbed the entrance ladder, their boots clanging against the rungs. Above them, the overhead hatches began to groan open, the sound a low, mechanical roar as the bay depressurized. Charlie, tethered to the craft's belly, worked the hold-down clamps with precise, robotic efficiency, his red eyes glowing in the dim light.

The air rushed out, a violent whoosh that tugged at Mia's clothes and lifted her and Callum upward, their bodies caught in the sudden vacuum. Her ears popped, her vision blurring as the pressure dropped. She clung to the ladder, her fingers aching, and reached for Callum, ensuring he didn't slip away.

Just as they reached the entrance hatch, Devory's strong hands grabbed them, hauling them into the Landing Craft's airlock. "Got you!" he shouted over the deafening rush of escaping air. He slammed the outer hatch shut, the clang reverberating through the small chamber, and hit the repressurization controls. Air hissed back in, the pressure stabilizing, though Mia's head still spun from the ordeal.

"Move!" Devory said, shoving them toward the acceleration couches lining the craft's interior. Mia stumbled, her legs like jelly, but she made it to her seat, strapping in with shaking hands. Callum collapsed into the couch beside her, his face ashen, staring blankly ahead. Aldo was already in the pilot's seat, his fingers flying over the controls, while Anna and Dr. Chen secured themselves nearby, their expressions grim.

Charlie's voice crackled over the internal comms. "Clamps releasing in three, two one—"

Aldo glanced back, his jaw tight. "Hold on. This is gonna be rough."

The Landing Craft shuddered as the *GenStar*'s centrifugal force flung it free, the scaffolding snapping away like brittle twigs. Mia's stomach lurched as the craft spun, the stars outside the viewport blurring into streaks. Aldo fired the thrusters, stabilizing their orientation, the low rumble vibrating through the hull.

Mia's eyes flicked to the countdown timer on the control panel—less than thirty seconds remained. Her heart pounded, the weight of Tiber's data stick, Rivel's sacrifice, and the *GenStar*'s doomed crew pressing against her. She reached for Callum's hand, squeezing it tightly. He didn't respond, but his fingers twitched, a faint acknowledgment.

"Ten seconds," Aldo called, his voice steady despite the stakes. "Brace for main engine burn."

The craft trembled as the engines roared to life, the acceleration pinning Mia to her couch. The *GenStar* shrank in the rear viewport, a massive, dying relic against the void of space. Then, a blinding flash erupted from its core, a silent explosion that consumed the ship in a sphere of white-hot light.

Mia closed her eyes, tears slipping down her cheeks. The *GenStar* was gone. Tiber, Kendricks, the Shard—all of it reduced to ash and memory.

But they were alive. Against all odds, they'd made it.

Devory's voice broke the silence, roaring with disbelief, "We're alive!"

Mia opened her eyes, her father's medal heavy in her pocket, Tiber's data stick pressing against her chest. The future was uncertain, but Tiber's words echoed in her mind: *You're the future, Mia. Tell your children, and your children's children.*

She would. For him, for her father, for Rivel, for all of them.